TELECOURSE FACULTY GUIDE FOR

UNSEEN LIFE ON EARTH

AN INTRODUCTION TO MICROBIOLOGY

CHRISTINE L. CASE

Skyline College

COORDINATED WITH

MICROBIOLOGY

AN INTRODUCTION

SIXTH EDITION

TORTORA • FUNKE • CASE

Unseen Life on Earth: An Introduction to Microbiology is produced by the American Society for Microbiology, Oregon Public Broadcasting, and Baker & Simon Associates with major funding from the Annenberg/CPB Project. *Unseen Life on Earth* is one component of the Microbial Literacy Collaborative, a partnership of organizations dedicated to raising public understanding of science and demonstrating the significance of microbes in our lives.

An imprint of Addison Wesley Longman, Inc.

San Francisco • Reading, Massachusetts
New York • Harlow, England • Don Mills, Ontario
Sydney • Mexico City • Madrid • Amsterdam

Publisher: Daryl Fox
Sponsoring Editor: Amy Folsom
Publishing Associate: Peggy Hammett
Marketing Manager: Lauren Harp
Production Editor: Larry Olsen
Production Service: Brian Jones
Composition and Text Design: Richard Kharibian
Artist: Shirley Bortoli
Cover Design: Yvo Riezebos
Cover Photo: SPL/Photo Researchers

ISBN 0–8053–2178–0

1 2 3 4 5 6 7 8 9 10—CRS—03 02 01 00 99

Credits:
Figures 2.1a, 2.1b, 2.2, 3c, 3d, 3e, 3f, 3g, 3h, 3.1b, 3.2, 4a, 4b, 5.2, 5.3, 7.1, 8.1, 8.2, 9.1, 9.2, 9.3, 10.1, 11.1, and 12.1 from *Microbiology: An Introduction,* Sixth Edition, by Tortora, Funke, and Case. Copyright © 1998 by The Benjamin/Cummings Publishing Company, Inc., an imprint of Addison Wesley Longman, Inc.

Addison Wesley Longman, Inc.
1301 Sansome Street
San Francisco, California 94111

ABOUT THE AUTHOR

Christine L. Case is a registered microbiologist and a professor of microbiology at Skyline College in San Bruno, California, where she has taught for the past 28 years. She teaches microbiology courses at a variety of levels, including biology majors, nonscience majors, allied health, and biotechnology professional-level courses. She earned her M.A. in Microbiology from San Francisco State University and her Ed.D. in Curriculum and Instruction from Nova Southeastern University. She has worked as a microbiologist with the U.S.D.A. Western Regional Research Lab and is a Director of the Society for Industrial Microbiology. Professor Case maintains a personal and professional commitment to the role of science in society. She has received the excellence in teaching award from the Northern California Chapter of the American Society for Microbiology and the Girl Scout award for contributions to science education.

To Don Biederman, with love, for his unflagging patience and encouragement.

ARRANGEMENTS FOR USE

Unseen Life on Earth: An Introduction to Microbiology may be used as a video resource for classes, libraries, and media centers or as a telecourse for distant learners. For further information, please call the phone numbers listed below. Colleges, universities, and other educational institutions may purchase the programs on videocassette or license the telecourse.

Purchase the Programs on Videocassette

Purchase the programs on videocassette for use in the classroom or media center. For more information, call 1-800-LEARNER or visit the Annenberg/CPB Project's Web site at http://www.learner.org and search on "Unseen Life on Earth."

Curriculum Package: $499

- 12 half-hour video programs on 12 cassettes
- *Telecourse Faculty Guide*
- Right to duplicate one set of videos

BONUS FOR ORDERING NOW: For a limited time only, if you order the *Unseen Life on Earth* curriculum package, you will also receive the *Intimate Strangers* prime-time video series (4 one-hour programs on 2 cassettes).

ASM MEMBER DISCOUNT: ASM members receive a 10% discount. Provide your membership number when you place your order.

For information on additional video duplication rights, call 1-800-LEARNER.

License the Telecourse

License the use of *Unseen Life on Earth: An Introduction to Microbiology* as a complete college credit telecourse for distant learners beginning in January 2000, or acquire an off-air or off-satellite taping license through the PBS Adult Learning Service. For more information, call 1-800-257-2578 or visit Adult Learning Service Online at http://www.pbs.org/als.

- Telecourse License: $500 per semester plus $20 per student
- Off-Air Taping License: $150 per series
- Off-Satellite Taping License: $250 standard fee; $150 for Adult Learning Satellite Service Associates

NOTE: *Intimate Strangers* is not part of the telecourse and will not be fed by the PBS Adult Learning Service. If you are offering the telecourse and wish to use *Intimate Strangers,* call 1-800-LEARNER to order those programs on videocassette.

CONTENTS

INTRODUCTION

The goal of this *Telecourse Faculty Guide* is to provide a framework for developing your distance learning course. Distance learning provides an unprecedented opportunity for students to take courses they might otherwise miss because of time constraints or travel requirements. A student's enthusiasm for learning is stimulated by your interest, and distance learning also provides a unique opportunity for you, the instructor, to interact with individual students.

How You Can Use *Unseen Life on Earth*

The video series and accompanying textbook can be used as a college-credit telecourse for distant learners in any of the following areas:

- For the lecture component of general microbiology or allied-health microbiology (coupled with a laboratory, this telecourse will fulfill the introductory microbiology requirement for these students).
- To fulfill a general education nonlab life-science requirement for nonbiology majors.
- To partially fulfill the life-science teaching certificate requirement for precollege teachers.

The video series can also be used for enrichment in the following ways:

- As a classroom resource for courses in general biology, health science, biotechnology, evolution, and ecology.
- As an education resource for bioscience businesses to train staff.
- As a resource for teacher in-service programs in the biological sciences.
- As discussion materials for high-school classes in biology and general science.
- As a video reference for public and university libraries and media centers.

Distance Learning

The video series features microbiologists demonstrating key principles and talking about their work to illustrate practical applications of microbiology. The personal profiles coupled with video microscopy and on-site visits to labs and hot springs provide dynamic "lectures." However, each video is a little less than 30 minutes long, about half the length of a normal class period, and the videos cannot allow for questions and discussion. The *Telecourse Study Guide* was designed as a guide to provide focus and reinforcement for important and difficult concepts. Exercises in the *Telecourse Study Guide* reinforce key topics and direct students to additional study opportunities in their textbook.

Your students will need you throughout the course. You might want to provide students with a schedule of dates and times that you will read and respond to their e-mailed questions. Consider setting up a listserve or electronic bulletin board so that students can interact with each other. Discussion questions for each unit can be posted on the bulletin board, and the students' responses can be used as the basis for a participation grade. More important, their responses will help you know whether they are studying appropriately.

If you cannot create a listserve, a class list with e-mail addresses should be distributed, and you should encourage students to form virtual study groups.

Tell the students how much study time this telecourse will require. The assumption for a traditional course is that each hour in class requires two hours outside of class; this totals nine hours a week for a course that meets three hours each week. The students will need to be told what is expected of them.

How to Use This Book

The *Telecourse Faculty Guide* contains the same material found in the *Telecourse Study Guide,* but it also contains some additional information for instructors, which is printed in color. A complete spectrum of multimedia learning resources for this telecourse is available. The resources are identified throughout the *Telecourse Study Guide* with the icons shown below.

This video series was designed for general education, general microbiology, and allied-health microbiology, so the depth of coverage and emphasis for each unit will vary, depending on your course. Here are some helpful suggestions.

Learning Objectives

The objectives from the assigned reading in *Microbiology: An Introduction*, Sixth Edition, are also listed in the *Telecourse Study Guide.* The objectives focus the student's attention on major concepts presented in the textbook. As in a regular lecture, all the objectives in a particular chapter cannot be covered in a video program. The objectives that are explicitly addressed in each video program are identified in this *Telecourse Faculty Guide* with a colored star (★).

You can choose the Learning Objectives for each unit that are necessary for your course. Your list can be provided to students in a printed syllabus or with weekly postings to your course Web site.

Reading

Students are asked to read the assigned pages in the textbook.

Key Terms and Concepts

Each unit includes a list of Key Terms and Concepts taken from the boldfaced terms in the assigned reading in *Microbiology: An Introduction.* As with the Learning Objectives, you can amend this list to suit the goals of your class.

Introduction

The Introduction is an overview of the main concepts covered in the textbook and the video program.

Preview of Video Program

The Preview is a capsule description of the video program. The Preview describes topics in the same order as they are shown in the video program.

Video Questions

Students do not have answers to the Video Questions, so you can use them for class discussion or graded assignments. If you do not formally use these questions, students may want you to post the answers for their reference.

Exercises

For each unit, there is a variety of review questions and exercises, usually including Concept Maps, Figures to label and answer questions about, Definitions, Matching Questions, and Short-Answer Questions. Answers to the kinds of questions listed above are provided for the student at the end of each unit.

Hypothesis Testing

Hypothesis Testing questions are designed to promote critical thinking and give you an opportunity to engage students in class discussions.

Questions from Other Resources

A list of Study Questions from the text is also included, as well as appropriate

CD Activities and

Web Activities.

Laboratory Requirement

The American Society for Microbiology's (ASM) Laboratory Core Curriculum (p. xiv of this *Telecourse Faculty Guide*) is considered essential to teach in every introductory microbiology laboratory, regardless of its emphasis. But how will you work a laboratory requirement into a telecourse? In this *Telecourse Faculty Guide,* lab experiments from Johnson/Case, *Laboratory Experiments in Microbiology,* Fifth Edition, are suggested for each unit. The virtual experiments on the CDs that accompany the text are designed to reinforce concepts, *not* to replace the actual laboratory experience. The laboratory for your course could be scheduled as a regular laboratory, perhaps in the evening or on weekends or as a separate course between semesters or during summer session.

THE VIDEO PROGRAMS

The *Telecourse Study Guide* and *Telecourse Faculty Guide* have been organized into 12 units to correspond with the 12 video programs. *The Unseen Life on Earth* video programs were developed around ASM's five core themes for beginning microbiology:

- Microbial cell biology
- Microbial genetics
- Microbial diversity and evolution
- Microorganisms in the environment
- Microorganisms and humans

The following are short descriptions of each of the 12 half-hour video programs.

Microbial Cell Biology

Unit 1: The Microbial Universe. The world of microorganisms is a dynamic one, and all other lifeforms depend on microbial metabolic activity. Students see how new microbes are discovered using culture techniques and genetic analysis.

Unit 2: The Unity of Living Systems. All cellular organisms—both prokaryotic and eukaryotic—share basic chemical similarities. Out of these similarities, however, emerge diverse patterns of cell assembly. Students encounter the tools they need to understand various cell types and their relationship to noncell entities such as viruses.

Unit 3: Metabolism. The metabolic pathways that produce energy create important environmental transformations. Although living organisms have diverse ways of meeting their energy needs, there is an amazing similarity among all lifeforms in how they carry out metabolism directed to the construction and use of necessary biological molecules.

Microbial Genetics

Unit 4: Reading the Code of Life. DNA is central to cell activity, carrying the information for all proteins and replicating with great fidelity—except in the important case of mutations. Organisms also regulate the products made from genes in an effort to conserve energy and adapt to new environments.

Unit 5: Genetic Transfer. Microbial populations achieve genetic diversity through horizontal gene transfer. Bacteria may transfer genes from one to another by conjugation, transformation, or transduction. Scientists often exploit these processes through recombinant DNA.

Microbial Diversity and Evolution

Unit 6: Microbial Evolution. Recent genetic techniques have led to new theories of evolution and the relationships among organisms. Students examine this "evolution revolution" using molecular sequences to trace the phylogenetic relationships of microbial life. Both the big picture of microbial evolution and the methods necessary for determining molecular phylogenies are examined.

Unit 7: Microbial Diversity. What is the relationship between the bacteria, archaea, and eukaryote branches of the tree of life, with their startling variety of organisms? Students see comparisons of organisms in their natural habitats and examine ways of studying these organisms both in those habitats and in the laboratory.

Microorganisms in the Environment

Unit 8: Microbial Ecology. Humans and all other lifeforms depend on microorganisms as the essential processors of oxygen, mineral nutrients for plant growth, and waste materials. Here we investigate some of the important environments dominated by microbes and how their presence is essential for human life.

Unit 9: Microbial Control. In certain situations, microbial control is a necessity. For instance, our food system requires sanitary conditions, and hospitals require sterilization techniques. Here we see the options available for various levels of microbial control.

Microorganisms and Humans

Unit 10: Microbial Interactions. There are many symbiotic relationships among microbes and between microbes and higher organisms. Students will examine fundamental examples of such relationships.

Unit 11: Human Defenses. Both nonspecific and specific defense strategies can defeat the invasion of microbial pathogens. Here students learn about the coordinated defense system of humans through visual analogy, animation, and examples of specific diseases.

Unit 12: Microbes and Human Disease. How microbes come into contact with humans and the many factors leading to disease outbreaks around the globe are examined here. Students learn about current efforts to track infectious diseases and the considerations necessary to control disease worldwide.

ASM CURRICULUM RECOMMENDATIONS

Since 1994, instructors from community colleges and four-year institutions across the country have worked to define the common guidelines for all introductory microbiology courses. Included in this definition are a recommendation and endorsement of a required laboratory experience. This inclusion of a laboratory experience as an integral part of all microbiology courses has been reaffirmed each year at the annual undergraduate microbiology conference sponsored by ASM. The annual conferences have fostered teaching practices to enhance learning based on these guidelines and have led to the development of curriculum materials, including the telecourse *Unseen Life on Earth: An Introduction to Microbiology.*

ASM recommends the following core curriculum guidelines for all introductory microbiology courses. These guidelines include prerequisite courses, core themes and concepts, and a laboratory core curriculum. The telecourse *Unseen Life on Earth: An Introduction to Microbiology* supports these curriculum recommendations. Updates to the curriculum recommendations can be found at http://www.asmusa.org.

Microbiology: An Introduction, Sixth Edition, by Tortora, Funke, and Case, and *Laboratory Experiments in Microbiology,* Fifth Edition, by Johnson and Case, also support all of ASM's curriculum recommendations.

PREREQUISITE COURSES

Introductory microbiology courses assume that students have acquired essential knowledge and skills from introductory biology and introductory chemistry courses at the undergraduate level.

CORE THEMES AND CONCEPTS

The core themes and concepts, considered essential to teach in every introductory microbiology laboratory regardless of its emphasis, include five overarching themes and 20 essential concepts. Faculty might add items appropriate to allied-health, applied, environmental, or microbiology-major courses.

The core themes and concepts are not meant to be a syllabus or outline. Rather, these themes and concepts are meant to support the development of learning objectives that can be met within the introductory microbiology course.

In these recommendations, the term *microbes* refers to all microorganisms, whether they be subcellular viruses and other infectious agents or cellular parts, including all prokaryotic and eukaryotic microbes. The asterisks in the list below denote those themes and concepts considered essential to the laboratory content.

	Chapter(s) in *Microbiology: An Introxduction,* Sixth Edition	Exercise(s) in *Laboratory Experiments in Microbiology,* Fifth Edition
Theme 1: Microbial cell biology*		
1. Information flow within a cell	8	27–28
2. Regulation of cellular activities	8	
3. Cellular structure and function*	4	3–7
4. Growth and division*	6, 28	20
5. Cell energy metabolism*	5	13–17
Theme 2: Microbial genetics*		27–31
1. Inheritance of genetic information	8	
2. Causes, consequences, and uses of mutations*	8–9	
3. Exchange and acquisition of genetic information	8	
Theme 3: Interactions and impact of microorganisms and humans*		
1. Host defense mechanisms	16–19	41–44
2. Microbial pathogenicity mechanisms*	15, 21–26	39–40, 45–49
3. Disease transmission	14	39–40
4. Antibiotics and chemotherapy*	7, 20	24–25
5. Genetic engineering	9	28, 30
6. Biotechnology	9, 28	30, 54, 56
Theme 4: Interactions and impact of microorganisms in the environment*		51–56
1. Adaptation and natural selection	8	
2. Symbiosis	27	
3. Microbial recycling of resources	27	
4. Microbes transforming the environment	6, 27–28	
Theme 5: Integrating themes*		
1. Microbial evolution	10	
2. Microbial diversity*	11–13, 27	32–38, 55

Laboratory Core Curriculum

Each element of the recommended laboratory core curriculum is covered in exercises in *Laboratory Experiments in Microbiology,* Fifth Edition.

The laboratory core curriculum, considered essential to teach in every introductory microbiology laboratory regardless of its emphasis, includes content themes (the sections above marked with an asterisk), laboratory skills, laboratory thinking skills, and laboratory safety skills. Faculty might add items appropriate to allied-health, applied, environmental, or microbiology-major courses.

The laboratory core curriculum is not meant to be a syllabus or outline. Rather, these core skills and topics are meant to support the development of learning objectives that can be met within the introductory microbiology laboratory. Depending on the specific emphasis of a course, a single laboratory session could meet multiple core objectives, focus on one objective, or emphasize a topic that is not in the laboratory core curriculum but is important to that particular course.

Laboratory Skills

The laboratory skills recommended by ASM are used throughout *Laboratory Experiments in Microbiology,* Fifth Edition. The exercise in which each skill is introduced is identified.

A student successfully completing basic microbiology will demonstrate the ability to do the following:

	Exercise
1. **Use a brightfield light microscope** to view and interpret slides, including	1
a. correctly setting up and focusing the microscope	
b. properly handling, cleaning, and storing the microscope	
c. correctly using all lenses	
d. recording microscopic observations	
2. **Properly prepare slides** for microbiological examination, including	
a. cleaning and disposing of slides	3
b. preparing smears from solid and liquid cultures	3
c. performing wet-mount and/or hanging-drop preparations	2
d. performing Gram stains	5
3. **Properly use aseptic techniques** for the transfer and handling of microorganisms and instruments, including	
a. sterilizing and maintaining the sterility of transfer instruments	10
b. performing aseptic transfer	11
c. obtaining microbial samples	11
4. **Use appropriate microbiological media** and test systems, including	
a. isolating colonies and/or plaques	11
b. maintaining pure cultures	11
c. using biochemical test media	12
d. accurately recording macroscopic observations	9
5. **Estimate the number of microbes** in a sample using serial-dilution techniques, including	11
a. correctly choosing and using pipettes and pipetting devices	
b. correctly spreading diluted samples for counting	
c. estimating appropriate dilutions	
d. extrapolating plate counts to obtain the correct colony-forming units (CFU) or plaque-forming units (PFU) in the starting sample	

	Exercise
6. **Use standard microbiology laboratory equipment** correctly, including	
a. using the standard metric system for weights, lengths, diameters, and volumes	11
b. lighting and adjusting a laboratory burner	3
c. using an incubator	9

Laboratory Thinking Skills

The laboratory thinking skills recommended by ASM are used in all experiments in *Laboratory Experiments in Microbiology,* Fifth Edition.

A student successfully completing basic microbiology will demonstrate an increased skill level in the following areas:

1. **Cognitive processes,** including
   a. formulating a clear, answerable question
   b. developing a testable hypothesis
   c. predicting expected results
   d. following an experimental protocol
2. **Analysis skills,** including
   a. collecting and organizing data in a systematic fashion
   b. presenting data in an appropriate form (graphs, tables, figures, or descriptive paragraphs)
   c. assessing the validity of the data (including integrity and significance)
   d. drawing appropriate conclusions based on the results
3. **Communication skills,** including
   a. discussing and presenting lab results or findings in the laboratory
4. **Interpersonal and citizenry skills,** including
   a. working effectively in teams or groups so that the task, results, and analysis are shared
   b. effectively managing time and tasks to allow concurrent and/or overlapping tasks to be done simultaneously, by individuals and within a group
   c. integrating knowledge and making informed judgments about microbiology in everyday life

Laboratory Safety Skills

The laboratory safety skills recommended by ASM are listed on pp. 2–4 of *Laboratory Experiments in Microbiology*, Fifth Edition, and are also integrated into every lab exercise to provide students with repetition and an opportunity to master laboratory safety.

A student successfully completing basic microbiology will demonstrate an ability to explain and practice safe

1. **Microbiological procedures,** including
   a. reporting all spills and broken glassware to the instructor and receiving instructions for cleanup
   b. using methods for aseptic transfer
   c. minimizing or containing the production of aerosols and describing the hazards associated with aerosols
   d. washing hands prior to and following laboratories and at any time contamination is suspected
   e. using universal precautions with blood and other body fluids and following the requirements of the O.S.H.A. Bloodborne Pathogen Standard
   f. disinfecting lab benches and equipment prior to and at the conclusion of each lab session, using an appropriate disinfectant and allowing a suitable contact time
   g. identifying and disposing properly of different types of waste
   h. reading and signing a laboratory safety agreement indicating that the student has read and understands the safety rules of the laboratory
   i. following good laboratory practices, including returning materials to their proper locations, the proper care and handling of equipment, and keeping the benchtop clear of extraneous materials
2. **Protective procedures,** including
   a. tying long hair back, wearing personal protective equipment (eye protection, coats, gloves, closed shoes; glasses may be preferred to contact lenses), and using such equipment in appropriate situations
   b. always using appropriate pipetting devices and understanding that mouth pipetting is forbidden
   c. never eating or drinking in the laboratory
   d. never applying cosmetics, handling contact lenses, or placing objects (fingers, pencils, etc.) in the mouth or touching the face
3. **Emergency procedures,** including
   a. locating and properly using emergency equipment (eye wash stations, first aid kits, fire extinguishers, chemical safety showers, telephones, and emergency numbers)
   b. reporting all injuries immediately to the instructor
   c. following proper steps in the event of an emergency
4. **In addition, institutions** where microbiology laboratories are taught will
   a. train faculty and staff in proper waste stream management
   b. provide and maintain all necessary safety equipment and information resources
   c. train faculty, staff, and students in the use of safety equipment and procedures
   d. train faculty and staff in the use of Material Safety Data Sheets (MSDS)

ACKNOWLEDGMENTS

The creation of this new telecourse truly has been a collaborative partnership.

Project Sponsors

The Annenberg/CPB Project
The Annenberg/CPB Project, founded in 1981, and the Annenberg/CPB Math and Science Project, founded in 1991, use electronic communications media to improve education for all Americans. Through these new media, the Projects give Americans access to quality college-level courses, and help schools and communities improve their elementary and secondary math and science programs. For more information, visit http://www.learner.org.

American Society for Microbiology
The American Society for Microbiology, founded in 1899, is the premier scientific organization for the microbiological sciences. The Society offers membership, conferences, publications, curriculum guidelines, an on-line library of teaching resources, career information, fellowships, and travel grants. For more information, visit http://www.asmusa.org.

Oregon Public Broadcasting
Oregon Public Broadcasting is a premier public broadcasting station and the producer of award-winning television programs for national and international distribution. For more information, visit http://www.opb.org.

Microbial Literacy Collaborative
The Microbial Literacy Collaborative, led by the American Society for Microbiology, is a partnership of organizations dedicated to raising public understanding of science and to demonstrating the significance of microbes in our lives. The partners have joined together to provide resources for improving microbial literacy worldwide. For more information, visit http://www.microbeworld.org.

The production of *Unseen Life on Earth* was guided by a national academic advisory committee for the telecourse. The committee members reviewed the video programs as they were developed and provided valuable suggestions on content and resources for the video programs. The committee members also reviewed and commented on Benjamin/Cummings's *Telecourse Study Guide, Telecourse Faculty Guide,* and *Telecourse Preview Guide.* The committee members and their primary affiliations are listed on the facing page.

Advisory Committee

Frederic K. Pfaender, Chair
Professor of Environmental Microbiology
University of North Carolina
Chapel Hill, NC

Robert E. Benoit
Professor of Biology
Virginia Polytechnic Institute and State University
Blacksburg, VA

Linda E. Fisher
Associate Professor of Biology and Microbiology
University of Michigan–Dearborn
Dearborn, MI

Suzanne V. Kelly
Chair of Math and Science Division
Scottsdale Community College
Scottsdale, AZ

Michael T. Madigan
Professor of Microbiology
Southern Illinois University
Carbondale, IL

Kristine M. Snow
Chair of Science Department
Fox Valley Technical College
Appleton, WI

Also, special thanks to the following individuals:

Oregon Public Broadcasting Production Team
David J. Davis, Executive Producer
John A. Booth, Series Producer
Mark C. Dorgan
Nell M. Gladson
Jim H. Leinfelder
Jessica L. Martin

ASM Project Manager
Amy L. Chang
Director of Education and Training
American Society for Microbiology
Washington, DC

Microbial Literacy Collaborative
Cynthia A. Needham, Chair
President
ICAN Productions
Stowe, VT

Susan E. Kee
Project Manager
Microbial Literacy Collaborative
American Society for Microbiology
Washington, DC

The following guidelines are given to students on p. xix of the *Telecourse Study Guide.*

For the most systematic coverage of each unit, you should

1. Read the Learning Objectives.
2. Skim the assigned pages in your textbook by reading the headings and figure legends.
3. Read the Introduction.
4. Read the Preview to the Video Program and the Video Questions.
5. Watch the Video Program. You might want to tape it to watch it a second time as you answer the Video Questions.
6. Read the assigned pages in your textbook. Pay special attention to the Key Terms and Concepts in this *Telecourse Study Guide.* Remember, there is a Glossary at the end of the textbook and on the Student Tutorial CD.
7. Do the Exercises in this *Telecourse Study Guide* and the assigned Study Questions in your textbook.
8. Practice using what you have learned by taking the quiz on the CD.
9. Review any questions that you missed.
10. Test your knowledge and comprehension by taking the quiz on the Web site.
11. Review any questions that you missed.

Unit 1

THE MICROBIAL UNIVERSE

Life would not long remain possible in the absence of microbes.
—LOUIS PASTEUR, 1867

LEARNING OBJECTIVES

	1.	List several ways in which microbes affect our lives.	p. 2
	2.	Explain the importance of observations made by Hooke and van Leeuwenhoek.	p. 3
	3.	Compare the theories of spontaneous generation and biogenesis.	p. 3
	4.	Identify the contributions to microbiology made by Needham, Spallanzani, Virchow, and Pasteur.	p. 3
	5.	Identify the importance of Koch's postulates.	p. 7
	6.	Explain how Pasteur's work influenced Lister and Koch.	p. 7
★	7.	Identify the contributions to microbiology made by Ehrlich, Fleming, and Dubos.	p. 10
	8.	Define bacteriology, mycology, parasitology, immunology, virology, and microbial genetics.	p. 11
	9.	Recognize the system of scientific nomenclature that uses genus and specific epithet names.	p. 14
★	10.	List the major characteristics of bacteria, fungi, protists, plants, and animals.	p. 14
★	11.	Differentiate among the major groups of organisms studied in microbiology.	p. 15
★	12.	List at least four beneficial activities of microorganisms.	p. 17
	13.	List two examples of biotechnology that use genetic engineering and two examples that do not.	p. 18
★	14.	Define normal microbiota.	p. 19
★	15.	Define and describe several infectious diseases.	p. 19
	16.	Define emerging infectious disease.	p. 19

READING

Chapter 1 and pp. 168, 282–283 (Veterinary Microbiology).

SUGGESTED LABS FROM JOHNSON AND CASE, *LABORATORY EXPERIMENTS IN MICROBIOLOGY*, FIFTH EDITION

Exercise 1: Use and Care of the Microscope
Exercise 2: Examination of Living Microorganisms
Exercise 35: Phototrophs: Algae and Cyanobacteria

Key Terms and Concepts

algae p. 17
antibiotics p. 10
aseptic techniques p. 7
bacteria p. 15
biogenesis p. 6
bioremediation p. 18
biotechnology p. 18
chemotherapy p. 10
emerging infectious diseases p. 20
eukaryotes p. 15
fermentation p. 7
fungi p. 15
gene therapy p. 18
genetic engineering p. 14
genus p. 14
germ theory of disease p. 7
helminths p. 17
immunity p. 9
immunology p. 11
infectious disease p. 19
Koch's postulates p. 9
microbial ecology p. 17
microbial genetics p. 14
molecular biology p. 14
mycology p. 11
normal microbiota p. 19
parasitology p. 11
pasteurization p. 7
pathogenic p. 3
prokaryotes p. 15
protozoa p. 15
recombinant DNA technology p. 14
resistance p. 19
specific epithet p. 14
spontaneous generation p. 5
synthetic drugs p. 10
virology p. 12
viruses p. 17

Introduction

Be sure to read the Study Outline for Chapter 1, pp. 22–26.

Living things too small to be seen with the unaided eye are called microorganisms or microbes. Microorganisms include bacteria, fungi, algae, protozoa, and viruses. These microorganisms live in a wide range of environments including hot springs, ice-cold oceans, and in and on animals, including humans. Relatively few microorganisms cause disease, and some microbes that live in animals are essential for the animal's health. In fact, microorganisms, especially bacteria, are essential to the maintenance of life on Earth.

The history of microbiology (Chapter 1) is a truly wonderful place to start a study of microbiology for several reasons:

1. The early researchers used good science, and their work provides examples of the scientific method (pp. 3–9).
2. Many of the developments in microbiology are also milestones in human history during the nineteenth and twentieth centuries. For example:
   - The discovery of antibiotics was the first time infectious diseases could be cured (p. 10).
   - The use of handwashing and aseptic techniques could prevent infectious disease (p. 9).
   - Recombinant DNA technology expanded the production of pharmaceuticals and industrial chemicals (p. 14).
   - The use of vaccines meant that infectious disease could be prevented (p. 9).
   - Microbiology played a decisive role in World War I (p. 3).

Ecology is one of the unifying themes of the biological sciences because all organisms are linked together in food webs. A food web consists of the pathways along which chemicals and energy are transferred from one organism to the next. An ecosystem consists of all the organisms in a particular place, or habitat, as well as the physical conditions with which they interact. The entire portion of Earth that is inhabited by living organisms, called the biosphere, could be considered an ecosystem. However, most ecologists study smaller habitats, such as a river or even a particular tree species that supports other organisms. Microbial ecologists study the roles of microbes in ecosystems. Each living organism occupies an ecological niche—that is, has a particular food and energy source and time and place in which it is active. Some organisms, such as plants and cyanobacteria, are the primary producers of an ecosystem. They are called primary producers because they produce organic molecules for their own growth and for all other organisms. Animals are called consumers because they depend on the producers for food and energy; consumers eat the producers or their products. Fungi and most bacteria also get their energy from the producers. These fungi and bacteria are the decomposers and are largely responsible for recycling the chemical elements (e.g., nitrogen, sulfur, phosphorus, carbon, and oxygen) in organic molecules back to forms that plants can use. Remember that all bacteria are not decomposers—for example, cyanobacteria are primary producers. You can see cyanobacteria growing in ponds and on greenhouse walls. Microbial ecology will be the subject of Units 8 and 10.

The term *germ* is not used in science to refer to microorganisms. The scientific definition of *germ* is "a rapidly growing cell"; it is used to describe embryonic cells such as wheat *germ*, which is the embryo of the wheat plant. The use of germ in reference to a microbe is not specific enough to use in science and medicine. A physician treating a patient for an infectious disease must know whether the disease is caused by a virus (e.g., AIDS), bacterium (e.g., streptococcal sore throat), fungus (e.g., athlete's foot), or protozoan (e.g., malaria) because each group of organisms is susceptible to a different category of drugs, not one all-purpose germicide. The principles of microbial control will be covered in Unit 9.

Evolution is the other unifying theme in the biological sciences. Virtually every organism has DNA, and the DNA of different organisms can be compared. These comparisons have led to the identification of previously unidentified organisms as well as a further understanding of the relationships among all organisms. The mechanism of evolution, natural selection, has led to an astounding array of different microbes (Units 6 and 7).

VIDEO: THE MICROBIAL UNIVERSE

Terms

The following new terms are introduced in this video:

bacterioplankton	ecosystems	ribosomal RNA
colonies	nucleotides	spread plate
cyanobacteria	primary producers	

Preview of Video Program

You will see a variety of microorganisms in the first two minutes of this video program. These images were photographed through a microscope at magnifications ranging from 100× to 1000×. Note the smallest cells: these are bacteria photographed at 1000×.

This video introduces the microbial world emphasizing that microorganisms are ubiquitous—that is, they are everywhere life has been found. The metabolic abilities of microbes are also used to produce a variety of goods such as foods and medicines, and certain microbes are essential for the growth of animals.

The microbial world was discovered over 300 years ago when Antoni van Leeuwenhoek made and used a microscope to look at scrapings from his teeth. Draw each of the different types of microbes as you see them in the video:

Protozoa.

Algae. Note the protozoan swimming between the two algal filaments. (~4 min.)

Fungi. ~4 min. Fungal spores are being released from a sporangium (spore sac). ~13.5 min. Mold growing on bread shows the filamentous hyphae. Look carefully for the small black sporangia.

Bacteria. What shape are they? Bacilli

Viruses.

How small is small? (~5 min.) Bacteria on the tip of a pin are used to emphasize the small size of bacteria. The bacteria on the pin are bacilli, or rod-shaped. You will learn more about cell structure in Unit 2.

The work of Louis Pasteur in the second half of the nineteenth century ushered in the Golden Age of Microbiology, during which scientists demonstrated that some microbes cause disease and other microbes perform many essential functions for life on Earth. The role of microorganisms in human diseases will be covered in Unit 12. Many examples of the importance of microorganisms, especially bacteria, are given in the video. Microbes can be called the "keepers of the environment" because they decompose organic compounds to recycle the chemical elements for the producers.

Microbes are also called the "creators of the environment." The activities of the first microbes created an environment in which other organisms could grow. For example, cyanobacteria were among the first organisms, and they produced molecular oxygen (O_2) as a by-product of photosynthesis. This form of oxygen made it possible for animals and other microbes to grow.

You will see a micrograph of amoeba among bacteria in the video (~14 min).
Sketch the cells and label the amoeba and bacteria.
A bacterial cell is ~25 times smaller than an amoeba.

Marine microbiologist Steve Giovannoni studies bacterioplankton. These are bacteria that are free-floating in the ocean, rather than attached to rocks, shells, and algae. Bacteria cannot be identified by looking at them through a microscope. Traditionally, microbiologists isolate bacteria by spreading a bacteria-containing sample onto the surface of a solid culture medium in a petri dish. This is called a streak plate (p. 168 in your textbook). Bacteria are usually streaked across the surface of the medium with a wire inoculating loop. In the video, Giovannoni holds an inoculating loop in the flame of a Bunsen burner to sterilize the loop. The loop must be sterilized before handling the bacteria to ensure that contaminates from the air and countertop do not enter the culture. The loop is again sterilized after handling the bacteria to prevent contaminating the lab when the loop is set down. Giovannoni demonstrates an alternative way to inoculate a culture medium called a spread plate, in which the bacterial sample is spread on the culture medium using a sterile glass spreading rod.

Within a day or two after the medium has been inoculated, the bacteria reproduce enough to form a colony that is visible without a microscope. Bacteria from the colonies can then be identified and studied. You will see growth on three different streak plates. Draw the appearance of one of the plates:

What color are the colonies? yellow, red, orange
Circle one colony.

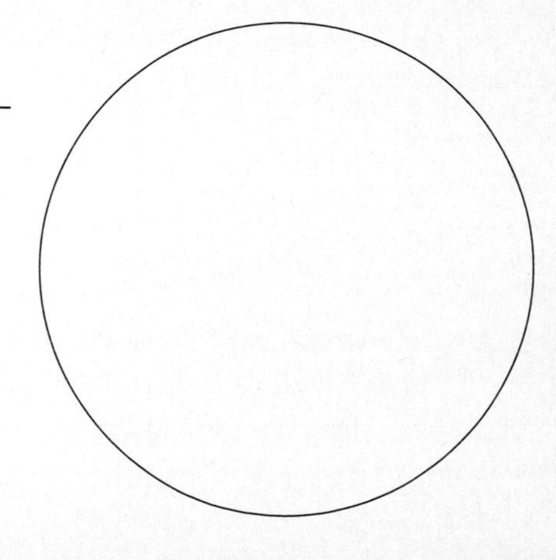

Microbiologists now realize that over 99% of the existing microbes have not been grown in laboratory cultures. Consequently, Giovannoni is using a new technique called PCR to identify bacteria that he hasn't grown. PCR will be described in Unit 7. Essentially, Giovannoni isolated the DNA from the microbial cells in a water sample, then located the specific genes within that DNA that encode ribosomal RNA (rRNA). All cells have ribosomes, so they all have genes for rRNA. Giovannoni identified bacteria from the Sargasso Sea by comparing their rRNA genes with databases of rRNA genes for known organisms.

$$\text{Water sample} \xrightarrow{\text{Lyse cells}} \text{DNA} \xrightarrow{\text{Copy DNA}}$$

$$\text{rRNA gene} \xrightarrow{\text{Compare to known organisms}} \text{Identification of organism}$$

He did find one organism he calls SAR11 that is unidentified and found in oceans and freshwater lakes around the world. Giovannoni describes the trial-and-error method used in his laboratory to try to grow SAR11. Even after identifying a bacterium, it is necessary to grow the bacterium in order to learn what nutrients it uses and what its role is in the ecosystem.

Video Questions

1. When did humans become aware of the existence of microorganisms? van Leeuwenhoek first saw microbes in 1673.

2. What is the ecological niche of most bacteria? Decomposer

3. Cellulose is the primary carbohydrate found in plants, and animals cannot digest cellulose. How can cows, bison, giraffes, moose, and termites live and grow by eating only plant matter? These animals use the by-products of cellulose-degrading bacteria in their digestive tract that break down the cellulose. This topic is covered in Unit 10.

4. Your body harbors how many times as many microbial cells as human cells? 10

5. What differentiates cyanobacteria from most other bacteria? Cyanobacteria use the same photophosphorylation that plants use, and they produce O_2.

6. Molecular oxygen (O_2) is produced by what types of organisms? 50% by plants and 50% by algae.

7. The following microorganisms are named in the video. Use your textbook to complete the table.

Microorganism	Bacterium, Fungus, Alga, Protozoan, or Virus	Normal Habitat
Diatoms	Alga	Surface water (e.g., lakes)
Enterococcus faecalis	Bacterium	Human intestines
Hantavirus	Virus	Rodents
Staphylococcus aureus	Bacterium	Human skin
Streptococcus pneumoniae	Bacterium	Human respiratory tract

EXERCISES

Concept Map 1.1

The lines on the map show the direction of chemicals and energy.

1. Label the lines that indicate energy.

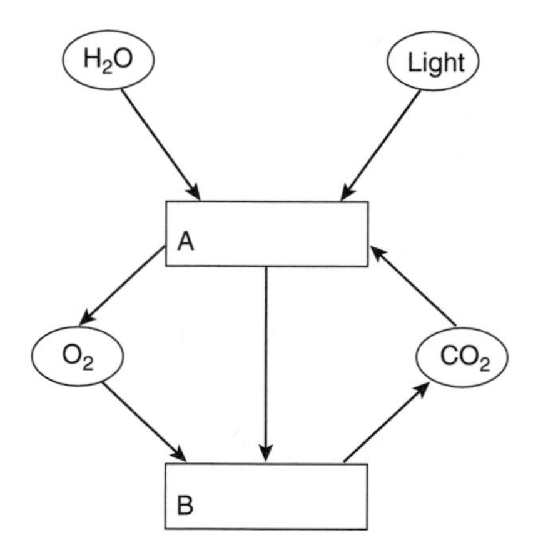

2. In which boxes do the following organisms belong?
   a. Plants
   b. Animals
   c. Cyanobacteria
   d. Most other bacteria
   e. Algae
   f. Bacterioplankton

Concept Map 1.2

Note that the lines with arrows show the direction of *energy* flow in a food chain.

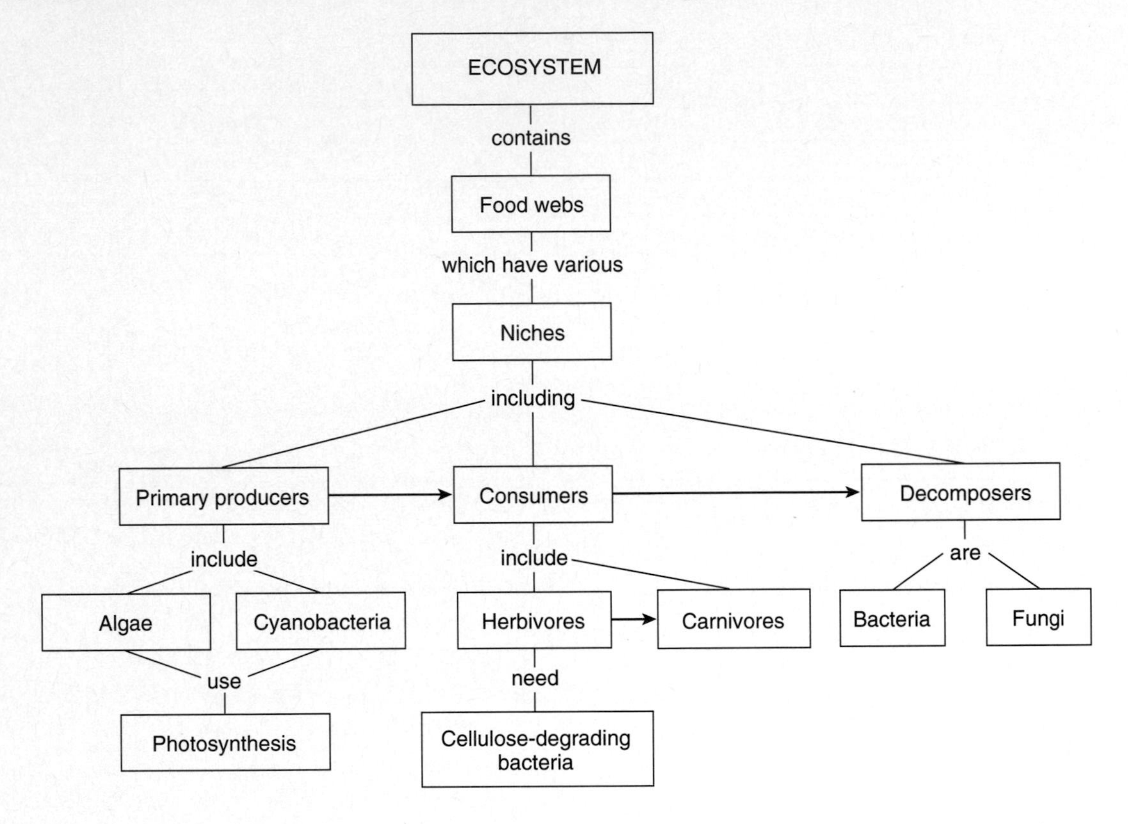

1. Which of the organisms are microbes?

2. What is the original energy source for this food chain?

3. Which group feeds at all energy levels?

4. Where do parasites such as pathogenic bacteria and viruses belong?

Figure 1.1

The scale shown in the figure is a logarithmic scale. An arithmetic scale would take several pieces of paper to cover this same range of sizes. See Appendix D in your textbook to review logarithms. Place the following in their correct place on this size line:

carbon atom (0.1 nm)
chicken egg (5 cm)
DNA molecule, width (2 nm)
E. coli, length (2 μm)
human egg (0.1 mm)
Human Immunodeficiency Virus (100 nm)
mitochondrion, length (2 μm)
nerve cell (1 m)
nucleus (6 μm)
red blood cell (7 μm)
ribosome (30 nm)
water molecule (0.4 nm)
your height (___ m)

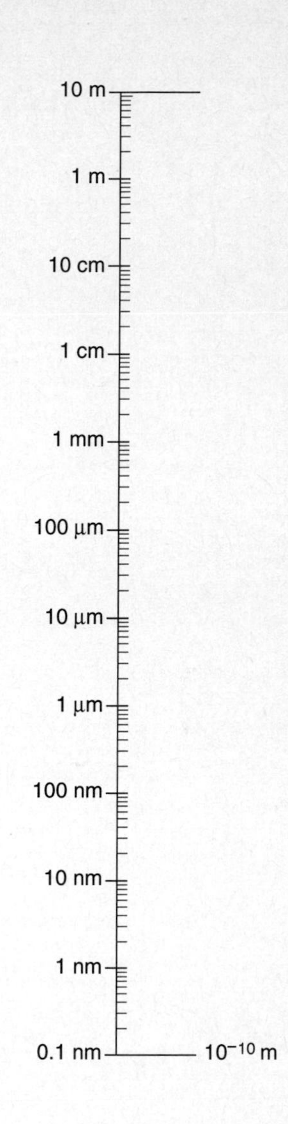

Definitions

Match the following statements to words from Key Terms and Concepts:

1. The eukaryotic organisms studied in microbiology. ____________________
2. Multicellular microorganisms. ____________________
3. Laboratory and clinical procedures employed to reduce contamination and the risk of infection. ____________________
4. The criteria that prove that a specific microorganism is the cause of a specific disease. ____________________
5. The idea that life comes from inanimate objects. ____________________
6. In the scientific name, *Escherichia coli*, *Escherichia* is the ____________________, and *coli* is the ____________________.
7. Antimicrobial chemicals produced by bacteria and fungi. ____________________
8. The use of microbes to clean up an oil spill. ____________________
9. The treatment of disease with a chemical. ____________________
10. Protection from a disease provided by vaccination. ____________________

History of Microbiology Questions

These questions match people who made significant contributions to the development of microbiology with their contributions. Match the following choices to the statements below.

a. Bassi
b. Crick
c. Ehrlich
d. Fleming
e. Hooke
f. Jenner
g. Koch
h. Lancefield
i. Lederberg
j. Linnaeus
k. Lister
l. Needham
m. Pasteur
n. Redi
o. Semmelweis
p. Spallanzani
q. van Leeuwenhoek
r. Watson

1. Proved the theory of biogenesis. ____
2. First to see microorganisms. ____
3. Developed the system of scientific nomenclature. ____
4. First to use the term *cells* to describe his microscopic observations. ____
5. First to suggest that a "magic bullet" could be used to treat infectious disease. ____
6. Used disinfection to prevent wound infections. ____
7. Proved that a specific microbe is the cause of a specific disease. ____
8. Showed that microbes cause chemical changes (e.g., fermentation) of food. ____
9. Showed that infections in obstetrical wards could be reduced by handwashing. ____
10. Devised a classification system for streptococci based on their cell walls. ____

Fill-In Questions

Use the following choices to complete the statements below. Choices may be used once, more than once, or not at all.

algae
bacillus
cell wall
coccus
cyanobacteria
fungi
helminths
nucleus
protozoa
spiral
viruses
yeasts

1. A rod-shaped bacterium is called a ____________________.
2. A spherical bacterium is called a ____________________.
3. ____________________ are photosynthetic prokaryotes.
4. ____________________ are photosynthetic eukaryotes.
5. ____________________ are microbes that are not composed of cells.
6. ____________________ are unicellular fungi.
7. ____________________ are multicellular animals that can cause disease.
8. Through a microscope, the difference between a yeast and a bacterium is ____________________.

Short-Answer Questions

1. List the organisms studied in microbiology.

2. Identify one commercial product available in stores that is made by bacteria.

3. Identify one agricultural use of bacteria.

4. Provide one example of a harmful activity of bacteria.

5. Where do most bacteria fit in the food chain?

6. Where do the cyanobacteria fit in the food chain?

7. Where do algae fit in the food chain?

8. What do fungi and humans have in common?

9. Name one emerging infectious disease that is *not* caused by a virus.

Hypothesis Testing

Students can post their experiments on an electronic bulletin board so that each student who responds has to add the next step or amend the previous step in the procedure.

Hypothesis: Microorganisms exist, and they are ubiquitous.

Experiment: Assuming that the microscope has not yet been invented, design an experiment to prove the existence of microbes. You might have to remind students to include the necessary controls and that they do not have microscopes.

Conclusion: Based on the experiments, do you accept or reject the hypothesis? Briefly explain.

Study Questions

Microbiology: An Introduction, Sixth Edition				
Pages	Review Questions	Multiple Choice Questions	Critical Thinking Questions	Clinical Applications Questions
24–26	All	All	All	All

CD Activity

Do the Chapter 1 quiz on the Microbiology Interactive Student Tutorial CD-ROM.

Web Activities

1. Do the Chapter 1 quiz on the Tortora/Funke/Case Online Course Companion Web site. (The URL is http://www.awlonline.com. You might want to bookmark this site for quick access during the remainder of this course.)
2. Read "The Microbiology of Chocolate" in the Applications section.
   a. When do the yeast begin to grow? Right away. The acetic acid bacteria? After yeast have produced ethyl alcohol, on which the acetic acid bacteria grow.

   b. Why don't they both grow at the same time? In this ecological succession, one organism changes the environment, which allows growth of a different organism.

ANSWERS

Concept Map 1.1

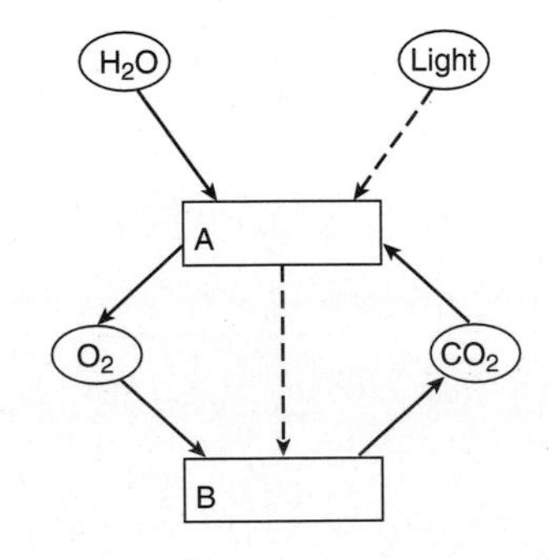

1. The broken (dashed) lines.
2. a. A
   b. B
   c. A
   d. B
   e. A
   f. B (supplying nutrients to the algae, per the video)

Concept Map 1.2

1. Algae, cyanobacteria, bacteria, fungi
2. Sunlight harnessed by the producers through photosynthesis
3. Decomposers and parasites
4. Parasites feed on each group, including the decomposers.

Figure 1.1

See Figure 3.2 (p. 59).

Definitions

1. fungi, algae, protozoa, helminths
2. fungi, algae, helminths
3. aseptic techniques
4. Koch's postulates or germ theory of disease
5. spontaneous generation
6. *Escherichia* is the genus, and *coli* is the specific epithet.
7. antibiotics
8. bioremediation
9. chemotherapy
10. immunity

History of Microbiology Questions

1. m
2. q
3. j
4. e
5. c
6. k
7. g
8. m
9. o
10. h

Fill-In Questions

1. bacillus
2. coccus
3. cyanobacteria
4. algae
5. viruses
6. yeasts
7. helminths
8. nucleus

Short-Answer Questions

1. Bacteria, fungi, algae, protozoa, viruses, and parasitic helminths.
2. Read labels and look for the name of an organism such as *Lactobacillus acidophilus* or the words *culture* or *fermented.*

3. *Rhizobium* bacteria are used by farmers to add nitrogen to the soil.
4. A disease such as pneumonia. Remember, this question asked for a *bacterial* disease, not a viral disease.
5. Bacteria are decomposers.
6. Cyanobacteria are producers.
7. Algae are producers.
8. Fungi and humans are both consumers and frequently use the same food. Look at the fungi growing on bread or an orange.
9. See Table 14.2 (p. 402) in your textbook.

Unit **2**

THE UNITY OF LIVING SYSTEMS

They [my mitochondria] are much less closely related to me than to each other and to the free-living bacteria out under the hill.

—LEWIS THOMAS, 1974

LEARNING OBJECTIVES

★ 1. Compare and contrast the overall cell structure of prokaryotes and eukaryotes. p. 76
2. Identify the three basic shapes of bacteria. p. 77
3. Describe the structure and function of the glycocalyx, flagella, axial filaments, fimbriae, and pili. p. 77
4. Compare and contrast the cell walls of gram-positive bacteria, gram-negative bacteria, archaea, and mycoplasma. p. 85
5. Describe the structure, chemistry, and functions of the prokaryotic plasma membrane. p. 88
6. Define simple diffusion, osmosis, facilitated diffusion, active transport, and group translocation. p. 91
★ 7. Identify the functions of the nuclear area, ribosomes, and inclusions. p. 95
8. Describe the functions of endospores, sporulation, and endospore germination. p. 96
9. Differentiate between prokaryotic and eukaryotic flagella. p. 98
10. Compare and contrast prokaryotic and eukaryotic cell walls and glycocalyxes. p. 100
11. Compare and contrast prokaryotic and eukaryotic plasma membranes. p. 101
12. Compare and contrast prokaryotic and eukaryotic cytoplasms. p. 101
★ 13. Define organelle. p. 101
14. Describe the functions of the nucleus, endoplasmic reticulum, ribosomes, Golgi complex, mitochondria, chloroplasts, lysosomes, and centrioles. p. 101
★ 15. Discuss evidence that supports the endosymbiotic theory of eukaryotic evolution. p. 106
★ 16. Differentiate between a virus and a bacterium. p. 360
★ 17. Describe the chemical composition and physical structure of both an enveloped and a nonenveloped virus. p. 361
★ 18. Compare and contrast the multiplication cycle of DNA- and RNA-containing animal viruses. p. 373
★ 19. Discuss some possible results of sequencing the human genome. p. 263

READING

Chapter 4 and pp. 61, 359–364, 373–377, 263 (Human Genome Project).

SUGGESTED LABS FROM JOHNSON AND CASE, *LABORATORY EXPERIMENTS IN MICROBIOLOGY*, FIFTH EDITION

Exercises 3–7: Staining Methods
Exercise 8: Morphological Unknown
Exercise 37: Isolation and Titration of Bacteriophages
Exercise 38: Plant Viruses

KEY TERMS AND CONCEPTS

The Prokaryotic Cell
active transport p. 93
amphitrichous p. 82
axial filaments p. 83
bacillus p. 77
bacterial chromosome p. 95
binary fission p. 76
capsule p. 77
carboxysomes p. 96
cell wall p. 85
chromatophores p. 91
coccobacilli p. 77
coccus p. 77
cytoplasm p. 94
diplococci p. 77
endoflagella p. 83
endospores p. 96
extracellular polysaccharide p. 77
facilitated diffusion p. 92
fimbriae p. 83
flagella p. 82
fluid mosaic model p. 90
gas vacuoles p. 96
germination p. 98
glycocalyx p. 80
group translocation p. 93
hypertonic solution p. 93
hypotonic solution p. 93
inclusions p. 95
isotonic solution p. 93
lipid inclusions p. 96
lophotrichous p. 82
lysis p. 85
magnetosomes p. 96
mesosomes p. 91
metachromatic granules p. 95
monomorphic p. 77
monotrichous p. 82
nucleoid p. 95
osmosis p. 92
osmotic lysis p. 85
osmotic pressure p. 92
peritrichous p. 82
pili p. 85
plasma membrane p. 85
plasmids p. 95
pleomorphic p. 77
polypeptides p. 85
polysaccharide granules p. 96
prokaryotes p. 76
protoplast p. 88
ribosomes p. 95
sarcinae p. 77
selective permeability p. 90
simple diffusion p. 91
slime layer p. 77
spheroplast p. 85
spiral p. 77
spirilla p. 77
spirochetes p. 77
sporulation p. 97
staphylococci p. 77
streptococci p. 77
sulfur granules p. 96
taxis p. 83
tetrads p. 77
thylakoids p. 91
vibrios p. 77
volutin p. 95

The Eukaryotic Cell
centrioles p. 106
chloroplast p. 105
chromatin p. 102
chromosomes p. 102
cilia p. 100
cytoplasm p. 101
cytoplasmic streaming p. 101
cytoskeleton p. 101
endocytosis p. 101
endoplasmic reticulum p. 102
endosymbiotic theory p. 106
eukaryotes p .76
flagella p. 100
glycocalyx p. 101
Golgi complex p. 103
histones p. 102
lysosomes p. 105
microtubules p. 100
mitochondria p. 103
nuclear envelope p. 102
nuclear pores p. 102
nucleoli p. 102
nucleus p. 101
organelles p. 101
plasma membrane p.101
ribosomes p. 103
rough ER p. 103
secretory vesicles p. 102
smooth ER p. 103
thylakoids p. 105
vacuole p. 105

Viruses
bacteriophages p. 360
capsid p. 362
capsomere p. 362
complex viruses p. 364
endocytosis p. 374
nonenveloped viruses p. 363
obligatory intracellular parasites p. 360
phages p. 360
spikes p. 363
uncoating p. 375
virion p. 361
viruses p. 360

INTRODUCTION

Be sure to read the Study Outlines for Chapter 4, pp. 107–109, and for Chapter 13, pp. 388–390.

The cell theory states that all living things are composed of cells. Cells are characterized by the ability to metabolize nutrients to generate energy and reproduce. There are a variety of energy-generating mechanisms, which will be discussed in Unit 3, and reproductive strategies vary; however, all cells possess a plasma membrane. The plasma membrane is composed of a phospholipid bilayer with proteins in and on it. The plasma membrane controls movement of materials into and out of the cell.

Stop for a moment and sketch a cell:

Does your sketch look like a fried egg? Remember that cells aren't flat, they are like the three-dimensional "little boxes" described by Robert Hooke. That "yolk" you drew represents the spherical or ovoid nucleus. Imagine that a cell looks like a basketball and the nucleus is held near the center by a protein network called the cytoskeleton. Your sketch represents a eukaryotic cell—that is, a cell that contains membrane-bound structures such as the nucleus; these structures are called organelles.

Other organelles include the mitochondria and (in plants) chloroplasts. All organelles except the mitochondria and chloroplasts are joined by a membrane called the endoplasmic reticulum to form a continuous membrane network in the cell.

In microbiology, we emphasize the study of prokaryotic cells because bacteria have prokaryotic cells and prokaryotic cells are not usually studied in other parts of the curriculum, such as anatomy and general biology. The structural and chemical differences between the prokaryotic and eukaryotic cell provide the basis for the action of antibiotics.

Prokaryotic cells do not have organelles. For example, the DNA that is housed in the eukaryotic nucleus is simply packed into the prokaryote's cytoplasm. Consequently, these cells can be smaller than eukaryotic cells. A prokaryotic cell contains one circular chromosome, whereas a eukaryotic cell has paired, linear chromosomes. A human cell has 46 chromosomes, or 23 pairs. The energy-generating metabolic processes of prokaryotes occur in the cytoplasm and plasma membrane rather than in organelles. Look back at your sketch of bacteria in Unit 1. You should not have drawn any structures inside the cell.

Both prokaryotic and eukaryotic cells may possess cell walls outside of the plasma membrane. Plants, fungi, and algae are eukaryotic organisms with cell walls. Archaea and bacteria consist of prokaryotic cells. Bacteria, formerly called Eubacteria, have walls composed of a unique chemical called peptidoglycan. The wall protects cells from osmotic lysis.

The structural simplicity of prokaryotic cells suggests that they evolved before eukaryotic cells. The endosymbiotic theory explains the evolution of eukaryotes from two prokaryotic cells that lived together. Compelling evidence for the endosymbiotic theory is provided by a comparison of a prokaryote with a mitochondrion and chloroplast. Read in your textbook about one bacterium, *Bdellovibrio*, that can live inside other bacteria (p. 61).

Viruses are not composed of cells. At their simplest, viruses are a few genes surrounded by a protein coat. They are obligate intracellular parasites, and when they are outside a host cell, they behave more like inanimate chemicals. Outside a host cell, viruses do not multiply and can be crystallized much like NaCl (table salt) in a solution. Inside a host cell, however, viral genes direct the cell's metabolic machinery to produce viruses instead of cell parts.

All cells have DNA and need energy in the form of ATP. The production of ATP is the topic of Unit 3, Metabolism.

VIDEO: THE UNITY OF LIVING SYSTEMS

Preview of Video Program

The video program begins with a description of prokaryotic and eukaryotic cells. The difference between these cell types is structural; both cells perform the same metabolic functions. However, the chemical reactions necessary for life take place in the cytoplasm and on the plasma membrane in a prokaryote and in organelles of eukaryotic cells.

The distinctive elements of prokaryotes (bacteria) are shown in the video. Many bacteria use flagella for motility. Bacterial flagella are structurally and chemically different from those of eukaryotes. All cells replicate by duplicating their DNA and dividing the cytoplasm around the new DNA. In bacteria, this process is called binary fission. The eukaryotic process is more complex because the linear chromosomes must be carefully segregated (during mitosis) so that each new cell receives a complete set of chromosomes.

Both types of cells have a plasma membrane that controls the entry and exit of materials to and from the cell and ribosomes for protein synthesis. The ribosomes in a prokaryote are slightly smaller than those of a eukaryote.

The video shows the ecosystem of a human large intestine, which houses a thriving population of bacteria including *Bacteroides*, *Fusobacterium*, and *Escherichia coli*. The bacterial population stays relatively constant, with the number of microbes dying (or being excreted) equal to the number that are dividing. These bacteria are essential for our health and well-being. One vital function is that they prevent other bacteria, which might cause disease, from colonizing.

The video introduces viruses, which are not composed of cells. Viruses are compared to a baseball with its host cell as the infield. Refer to the size line you made in Unit 1 to review the sizes of viruses and cells. Viral genes are made of DNA or RNA, never both. The genes are surrounded by a protein coat or capsid, not a plasma membrane; and the capsid may be surrounded by an envelope. A virus must use the ribosomes, amino acids, and energy of a host cell to reproduce itself. Viruses run into cells by accident, and the virus may attach if the cell has the appropriate receptor.

The structure and metabolism of cells and viruses are the products of their nucleic acids (DNA, or RNA for some viruses). In cells, the DNA is arranged in molecules called chromosomes. The entire chromosome set of an organism is called the genome. A bacterial genome consists of approximately 2,000 genes on one chromosome. The human genome is 60,000 to 80,000 genes on 24 chromosomes. Genomics, the molecular study of genomes, is introduced in an interview with Craig Venter, who is leading a project to sequence the human genome.

Venter says that "Each genome is the recorded history of life" because a cell's genes are passed down to its offspring. Many genes have been conserved—that is, passed unchanged from one generation to the next. Consequently, all human genes can be traced to genes that are found in the archaea and in the bacteria. Microbial evolution is the topic of Unit 6.

Video Questions

1. How can you distinguish between the prokaryotic and eukaryotic cells shown in the video? Characterize the bacteria shown in the video according to morphology: bacilli, cocci, or spirilla. Bacilli. Organelles are visible in the eukaryotic cells. The bacteria are smaller and lack organelles.

2. Why do the following provide evidence of evolution? The presence of DNA in cells. The presence of the same genes in bacteria and humans. The presence of DNA in all cells and the presence of the same genes in bacteria and humans suggest that all cells came from a common ancestor that had DNA.

3. Which evolved first, a virus or its host cell? The host cell most likely evolved first, then the virus arose from some abnormal cell division or DNA replication event.

4. Are viruses alive? Students can discuss the characteristics viruses share with nonliving chemicals and with living cells.

5. The following statement is made in the video: "What microbes are ultimately telling us is that each of us contains within our genetic material a written history of life on Earth." Briefly explain the statement. From studying the DNA inside them, we have learned that cells like ours were formed over a billion years ago by a marriage between different single-celled microbes.

6. If genes were equally distributed on each chromosome, how many genes would each human chromosome carry? Each chromosome would carry approximately 3,333 genes if 80,000 genes were evenly distributed on 24 chromosomes.

7. Of what value to humankind is the human genome project? To help us understand how humans are similar to other animals and how they are different. Information about specific genes may help researchers develop treatments for or prevent inherited diseases.

EXERCISES

Concept Map 2.1

This map shows terms related to cell structure.

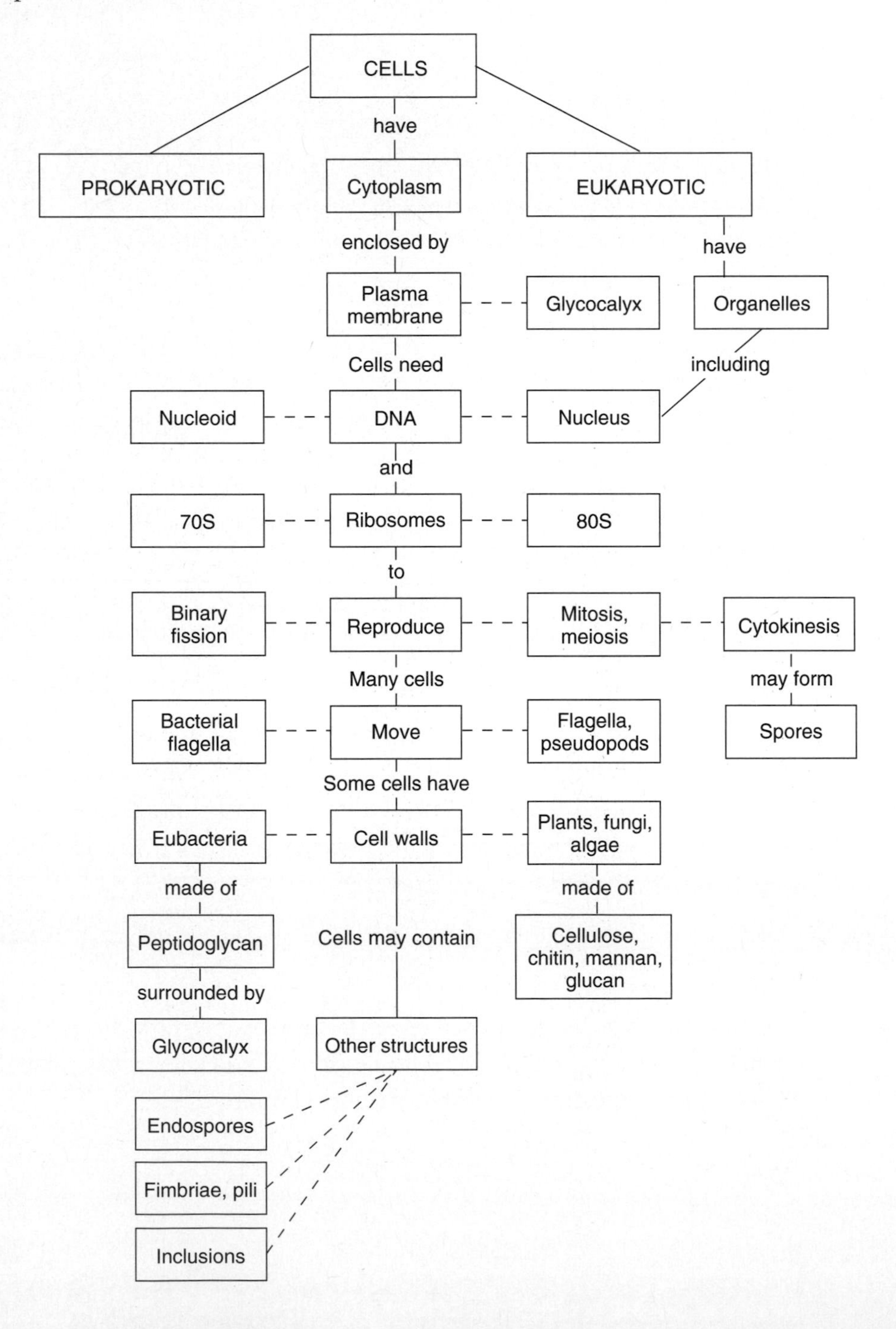

1. What are the minimum structural requirements for a cell?

2. Which kingdoms have eukaryotic cells? Which kingdoms have prokaryotic cells?

3. What structures are basically the same in both prokaryotes and eukaryotes?

4. What functions are basically the same in both prokaryotes and eukaryotes?

Figure 2.1

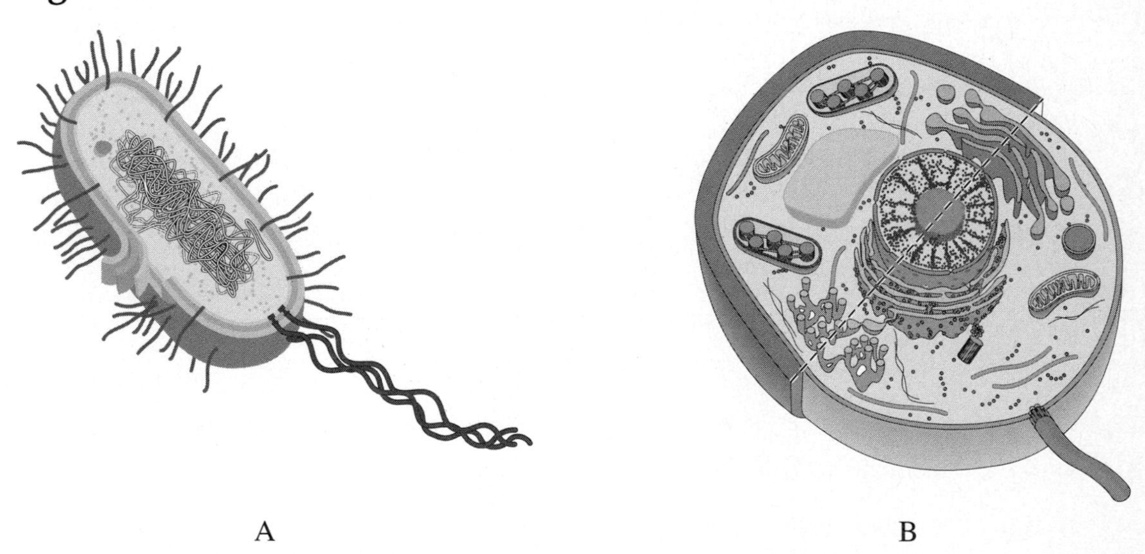

1. Identify which is the prokaryotic cell, and which is the eukaryotic cell.
2. Using different colors, color the 70S ribosomes, circular chromosomes, the membranes that contain the respiratory enzymes, and the thylakoid membranes in each cell.
3. How do plant and animal cells differ? How are they similar?

Figure 2.2

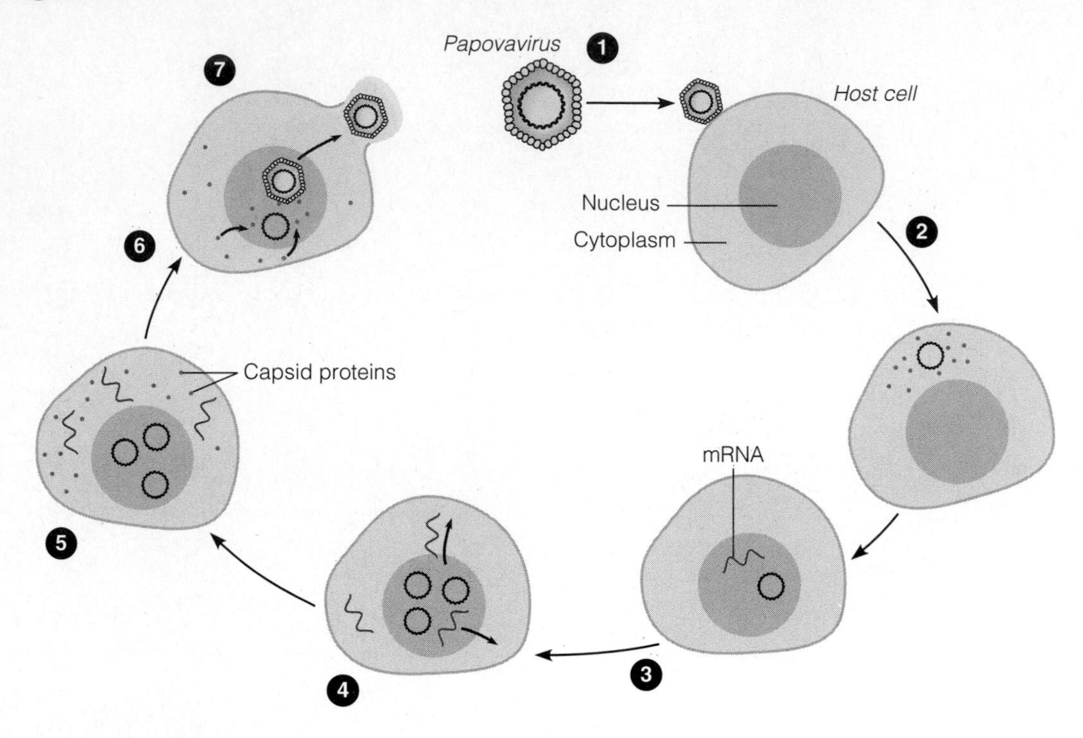

1. Label the viral DNA and capsid.
2. How does the structure of this virus differ from that of a hepadnavirus?

3. What diseases are caused by papovaviruses?

4. Identify the following steps in the figure:
   a. Attachment
   b. Penetration
   c. Protein synthesis
   d. DNA replication
   e. Release

Definitions

Match the following statements to words from Key Terms and Concepts:

1. Membranes that contain pigments for photosynthesis. ______________________
2. A gram-negative cell after treatment with lysozyme. ______________________

3. External proteins used for attachment. ______________________
4. Process by which water enters a cell. ______________________
5. Movement of a substance across a plasma membrane that requires energy and enzymes; the substance is not changed during movement. ______________________
6. The sites of protein synthesis in a cell. ______________________
7. The structure that holds cytoplasm. ______________________
8. Process by which most bacteria reproduce. ______________________
9. Highly resistant forms of bacterial cells. ______________________

Matching Questions

Match the following choices to the statements below:

a. Active transport
b. Axial filament
c. Capsule
d. Carboxysome
e. Cell wall
f. Chloroplast
g. Endospore
h. Facilitated diffusion
i. Fimbriae
j. Metachromatic granules
k. Microtubules
l. Mitochondrion
m. Mitosis
n. Nucleus
o. Osmosis
p. Passive diffusion
q. Peritrichous
r. Plasma membrane
s. Sex pilus

1. Passive processes by which material crosses the plasma membrane. ____
2. Phosphate-storage structure in prokaryotic cells. ____
3. Process by which carbon dioxide leaves an animal cell. ____
4. Organelle responsible for photosynthesis. ____
5. This structure may prevent white blood cells from phagocytizing bacteria. ____
6. These external proteins allow bacteria to stick to mucous membranes. ____
7. Term describing flagella present all over a bacterial cell. ____
8. Structure used for movement by a spirochete. ____
9. Eukaryotic flagella and cilia are composed of these. ____
10. The fluid mosaic model describes this structure. ____

Comparison Questions

For the following list of terms, mark

P if it is a prokaryotic structure or process
E if it is a structure or process that a eukaryote possesses
B if it occurs in both prokaryotes and eukaryotes
N if it is a structure or process that occurs in neither prokaryotes nor eukaryotes

___ 1. Cell membrane
___ 2. Retrovirus
___ 3. 80S ribosomes
___ 4. Cellulose cell wall
___ 5. Mitochondrion
___ 6. Peptidoglycan
___ 7. Nucleus bounded by a membrane
___ 8. Endospores
___ 9. Osmosis
___ 10. Microtubules
___ 11. Cilia

Fill-In Questions

Use the following choices to complete the statements below. Choices may be used once, more than once, or not at all.

attachment
capsid
cell wall
controls movement into cell
DNA
pinocytosis
phospholipids
plasma membrane
plasmolyze
prevent osmotic lysis
proteins
RNA
uncoating

1. Cell respiration and photosynthesis occur on the ____________________ in prokaryotic cells.
2. A virus enters a host cell by ____________________.
3. Once a virus is inside the host cell, viral nucleic acid is released from the capsid; this is called ____________________.
4. Viral genes can be made of ____________________ or ____________________.
5. A cell's genes are made of ____________________
6. A cell placed in a hypertonic environment will ____________________
7. Viral nucleic acid is surrounded by a ____________________.
8. The plasma membrane is made of ____________________.

9. The function of the cell wall is ______________________.
10. The outer covering on a human cell is the ______________________.

Short-Answer Questions

1. The following antibiotics are used to treat microbial infections. Why don't they harm the human host? Why don't they work against viral infections?

   a. Penicillin, inhibits peptidoglycan synthesis.

   b. Streptomycin, inhibits protein synthesis at 70S ribosomes.

   c. Sulfonamides, inhibits folic acid synthesis.

2. Differentiate between a gram-positive and a gram-negative cell wall.

3. If you can't identify bacteria from a Gram stain, why would a physician want Gram-stain results before prescribing treatment?

4. Name a species of endospore-forming bacteria. What is the importance of the endospore to this species?

5. *Streptomyces* bacteria form chains of cells and produce conidiospores for reproduction. *Penicillium* fungi form chains of cells and produce conidiospores for reproduction. Why is *Streptomyces* classified as a bacterium and not a fungus?

6. In 1997, researchers discovered a 1 mm microbe (*Thiomargarita namibiensis*) in ocean sediments. This microbe is visible without a microscope. What criteria did they use to classify *Thiomargarita* as a bacterium and not a protist?

7. Sex pili in bacteria are not for reproduction. What are they used for?

8. Differentiate between lysozyme and lysosome.

9. How do viruses differ from cells?

10. Viruses do not have any way to process food to get energy and raw materials. How, then, do viruses get energy and raw materials for reproduction?

Hypothesis Testing

Students can post their arguments/evidence on an electronic bulletin board to arrive at a class conclusion.

Hypothesis: Eukaryotic organelles arose from prokaryotes living inside a host prokaryotic cell.

Data/Observations: What evidence supports this hypothesis? Mitochondria and chloroplasts have a single, circular molecule of DNA and 70S ribosomes, and reproduce by binary fission.

Conclusion: From these data, do you accept or reject the hypothesis? Briefly explain.

Study Questions

***Microbiology: An Introduction,* Sixth Edition**				
Pages	**Review Questions**	**Multiple Choice Questions**	**Critical Thinking Questions**	**Clinical Applications Questions**
109–111	All	All	All	All
390–392	All	All	All	All

CD Activities

Do the Chapter 4 and Chapter 13 quizzes.

Web Activities

1. Do the Chapter 4 and Chapter 13 quizzes.
2. Read about the use of viruses to cure disease (Applications section). How can viruses be used to cure diseases? Hypothetically, viruses might be used to kill disease-causing bacteria. Viruses are used to make vaccines that can prevent disease.
3. Read about the discovery of magnetosomes (Applications section). How did Blakemore use scientific methodology to discover magnetosomes? After observing that the cells were moving in the same direction, he set up controlled experiments to determine the stimulus for the cells' movement.

ANSWERS

Concept Map 2.1

1. Plasma membrane, ribosomes, chromosome (DNA)
2. Plantae, Animalia, Fungi, and Protista have eukaryotic cells. Bacteria and Archaea have prokaryotic cells.
3. The plasma membrane is the same in both eukaryotes and prokaryotes. Ribosomes are similar in both cell types.
4. Both cell types must generate energy and reproduce.

Figure 2.1

1. The prokaryotic cell is A.
2. In the prokaryotic cell, you should have colored the 70S ribosomes, circular chromosomes, and plasma membrane. In the eukaryotic cell, you should have colored the mitochondria and chloroplasts, which contain 70S ribosomes, circular chromosomes, respiratory enzymes (mitochondria), and thylakoid membranes (chloroplasts).
3. Plant cells have a cell wall and chloroplasts; they lack lysosomes; the endoplasmic reticulum is smaller and plant cells do not phagocytize. Both plant and animal cells have plasma membranes, mitochondria, a Golgi complex, 80S ribosomes, and a nucleus.

Figure 2.2

1. Refer to Figure 13.16 in your textbook.
2. Hepadnavirus has an envelope and reverse transcriptase.
3. Warts, uterine cancer.
4. See Figure 13.16 in your textbook.

Definitions

1. thylakoids
2. protoplast
3. fimbriae
4. osmosis
5. active transport
6. ribosomes
7. plasma membrane
8. binary fission
9. endospores

Matching Questions

1. o, p, h
2. j
3. p
4. f
5. c
6. i
7. q
8. b
9. k
10. r

Comparison Questions

1. B
2. N
3. E
4. E
5. E
6. P
7. E
8. P
9. B
10. E
11. E

Fill-In Questions

1. plasma membrane
2. pinocytosis
3. uncoating
4. DNA; RNA
5. DNA
6. plasmolyze
7. capsid
8. phospholipids
9. prevent osmotic lysis
10. plasma membrane

Short-Answer Questions

1. a. Only bacteria have peptidoglycan cell walls.
   b. The cytoplasmic ribosomes in humans are 80S; viruses do not have ribosomes.
   c. Human cells and viruses cannot make folic acid.

2. A gram-positive cell wall is composed of a thick layer of peptidoglycan with teichoic acids. Gram-negative cell walls have a thin layer of peptidoglycan surrounded by a phosopholipid outer membrane.
3. An antibiotic may be indicated by the gram reaction of the pathogen. For example, penicillin can be effective against a gram-positive bacterium but, generally, will not affect a gram-negative bacterium.
4. You should provide a genus and species name and discuss survival of the species in an unfavorable environment. For example, *Clostridium botulinum* would need an endospore to survive in the presence of air.
5. *Streptomyces* is prokaryotic.
6. Seeing prokaryotic cell structure.
7. Pili are used for conjugation, in which one cell gives DNA to another cell. This is not reproduction because a third, or offspring, cell does not result from the exchange.
8. Lysozyme is an enzyme that hydrolyzes or digests peptidoglycan. Lysosome is a eukaryotic organelle that contains digestive enzymes.
9. Viruses are not enclosed by a plasma membrane and have no energy-generating metabolism.
10. The virus uses the host cell's energy and materials (e.g., amino acids and ATP).

Unit **3**

METABOLISM

The unity in the divergent metabolic processes of the microbes [is the same as] the metabolism of higher organisms.
—ALBERT JAN KLUYVER, 1924

LEARNING OBJECTIVES

	1.	Define metabolism, and describe the fundamental differences between anabolism and catabolism.	p. 112
	2.	Identify the role of ATP as an intermediate between catabolism and anabolism.	p. 112
	3.	Describe the mechanism of enzymatic action.	p. 113
	4.	List the factors that influence enzymatic activity.	p. 113
	5.	Define ribozyme.	p. 113
	6.	Explain what is meant by oxidation-reduction.	p. 119
	7.	List and provide examples of three types of phosphorylation reactions that generate ATP.	p. 119
★	8.	Explain the overall function of metabolic pathways.	p. 121
	9.	Describe the chemical reactions of glycolysis.	p. 122
	10.	Explain the products of the Krebs cycle.	p. 122
	11.	Describe the chemiosmotic model for ATP generation.	p. 122
★	12.	Compare and contrast aerobic and anaerobic respiration.	p. 130
★	13.	Describe the chemical reactions of, and list some products of, fermentation.	p. 132
	14.	Describe how lipids and proteins are prepared for glycolysis.	p. 134
	15.	Provide an example of the use of biochemical tests to identify bacteria.	p. 135
	16.	Compare and contrast cyclic and noncyclic photophosphorylation.	p. 136
★	17.	Compare and contrast the light and dark reactions of photosynthesis.	p. 136
	18.	Compare and contrast oxidative phosphorylation and photophosphorylation.	p. 136
★	19.	Categorize the various nutritional patterns among organisms according to carbon source and mechanisms of carbohydrate catabolism and ATP generation.	p. 139
	20.	Describe the major types of anabolism and their relationship to catabolism.	p. 142
	21.	Define amphibolic pathways.	p. 145
	22.	Compare primary, secondary, and tertiary sewage treatment.	p. 731
	23.	List some of the biochemical activities that take place in an anaerobic sludge digester.	p. 731
★	24.	Define activated sludge system, trickling filter, septic tank, and oxidation pond. (BOD is covered in Unit 8.)	p. 731
	25.	Name four beneficial activities of microorganisms in food production.	p. 746
	26.	Distinguish between chemically defined and complex media.	p. 162
★	27.	Justify the use of each of the following: anaerobic techniques, living host cells, candle jars, selective and differential media, enrichment media.	p. 162

28. Define colony. p. 167
29. Describe how pure cultures can be isolated by using the streak plate method. p. 167
★ 30. Define bacterial growth, including binary fission. p. 169
★ 31. Compare the phases of microbial growth, and describe their relation to generation time. p. 170

READING

Chapter 5 and pp. 4, 159, 170–171, 731–738, and 746–749.

To review general chemistry, pp. 27–34, organic chemistry, pp. 40–51.

SUGGESTED LABS FROM JOHNSON AND CASE, *LABORATORY EXPERIMENTS IN MICROBIOLOGY,* FIFTH EDITION

Exercises 13–18: Microbial Metabolism

Exercise 20: Determination of a Bacterial Growth Curve

Exercise 53: Microbes in Food: Contamination

Exercise 54: Microbes Used in the Production of Foods

KEY TERMS AND CONCEPTS

Metabolism

activation energy p. 113
active site p. 115
aerobe p. 125
aerobic respiration p. 125
alcohol fermentation p. 133
allosteric p. 118
allosteric site p. 118
amination p. 145
amphibolic pathways p. 145
anabolism p. 112
anaerobe p. 125
anaerobic respiration p. 125
anoxygenic p. 140
apoenzyme p. 114
ATP p. 120
autotrophs p. 140
Calvin-Benson cycle p. 139
carbohydrate catabolism p. 122
carbon fixation p. 136
catabolism p. 112
catalysts p. 113
chemiosmosis p. 128
chemoautotrophs p. 142
chemoheterotrophs p. 142
chemotrophs p. 140
coenzyme p. 114
coenzyme A p. 114
cofactor p. 114
competitive inhibitors p. 118
cyclic photophosphorylation p. 138
cytochromes p. 127
dark (light-independent) reactions p. 137
decarboxylation p. 126
dehydrogenation p. 120
denaturation p. 117
electron transport chain pp. 121, 127
end-product inhibition p. 118
Entner-Doudoroff pathway p. 125
enzymes p. 113
enzyme-substrate complex p. 115
FAD p. 114
feedback inhibition p. 118
fermentation p. 132
fermentation test p. 135
flavoproteins p. 127
FMN p. 114
glycolysis p. 123
green nonsulfur bacteria p. 142
green sulfur bacteria p. 140
heterolactic p. 134
heterotrophs p. 140
holoenzyme p. 114
homolactic p. 133
inhibition p. 118
Krebs cycle p. 125
lactic acid fermentation p. 132
light (light-dependent) reactions p. 137
lithotrophs p. 140
metabolism p. 112
metabolism pathways p. 113
NAD^+ p. 114
$NADP^+$ p. 114
noncompetitive inhibitors p. 118
noncyclic photophosphorylation p. 138

organotrophs p. 140
oxidation p. 120
oxidation-reduction p. 120
oxidative phosphorylation p. 121
oxygenic p. 140
parasites p. 142
pentose phosphate pathway p. 125
phosphorylation p. 121
photoautotrophs p. 140
photoheterotrophs p. 141
photophosphorylation p. 121
photosynthesis p. 136
phototrophs p. 140
purple nonsulfur bacteria p. 142
purple sulfur bacteria p. 141
redox reaction p. 120
reduction p. 120
respiration p. 125
ribozyme p. 119
saprophytes p. 142
saturation p. 117
substrate p. 113
substrate-level phosphorylation p. 121
transamination p. 145
turnover number p. 114
ubiquinones p. 127

Microbial Growth

agar p. 162
bacterial growth curve p. 170
binary fission p. 169
budding p. 169
capnophiles p. 165
chemically defined medium p. 163
colony p. 168
complex medium p. 163
culture p. 162
culture medium p. 162
death phase p. 171
deep-freezing p. 168
differential media p. 166
enrichment culture p. 166
exponential growth phase p. 171
generation time p. 169
inoculum p. 162
lag phase p. 170
logarithmic decline phase p. 171
log phase p. 171
lyophilization p. 168
nutrient agar p. 164
nutrient broth p. 164
selective medium p. 166
stationary phase p. 171
sterile p. 162
streak plate method p. 168

Wastewater Treatment

activated sludge system p. 733
anaerobic sludge digesters p. 735
primary sewage treatment p. 732
secondary sewage treatment p. 733
sludge p. 732
tertiary sewage treatment p. 737
trickling filters p. 734

INTRODUCTION

Be sure to read the Study Outlines for Chapter 5, pp. 147–151; Chapter 27, pp. 739–740; and Chapter 28, p. 758.

At first glance, cellular metabolism looks quite daunting. Don't be frightened or frustrated during your studies; metabolism is fascinating, and there are really just a few basic principles. The study of metabolism is important because the basic principles are the same for all organisms, including humans. By studying metabolism, scientists can identify and develop treatments for metabolic diseases, find microbes with unique metabolic pathways or enzymes that can be used to detoxify toxic wastes or produce useful products, and understand how different organisms interact in the ecosystem. A list of industrial uses of microbial metabolism is shown in Table 5.4 (p. 135) in your textbook.

Basically, all cells need to grow and reproduce. Building a house can provide an analogy to making a cell. To build a house, you need a blueprint of the plan, raw materials, and energy to put the raw materials together. In a cell, DNA provides the blueprint, the cell needs a carbon source to provide the materials for new cells, and the cell needs an energy source. DNA will be discussed in Unit 4. In this unit, we will look at how the carbon and energy sources are processed in the cell's metabolism. Metabolism is the sum of all of the chemical reactions that occur in a living cell. Metabolism consists of catabolism (energy-producing) and anabolism (energy-requiring). As shown in Figure 5.1 (p. 113), energy, usually in the form of ATP, is the intermediate that joins catabolic and anabolic reactions.

ATP is made by adding a phosphate ion to (phosphorylating) an ADP molecule. This takes 7.3 kcal, which is obtained from a catabolic pathway. The cell must have phosphate ions from the environment and have, or make, ADP:

$$ADP + PO_4^{3-} \xrightarrow{\text{Energy}} ATP$$

ATP is made by three methods, classified according to the energy source. In substrate-level phosphorylation, energy comes from a chemical bond holding a PO_4^{3-} to an organic molecule. In oxidative phosphorylation, the energy comes from a series of redox reactions. And, in photophosphorylation, light provides the energy.

When the cell needs energy, ATP molecules are hydrolyzed to liberate energy stored in the bond(s) holding the phosphate ion(s). ATP is discussed on pp. 51 and 120–121 in your textbook.

Carbon Sources

Producers such as plants and photosynthetic bacteria use an inorganic carbon source such as CO_2. These organisms are called autotrophs or lithotrophs. Consumers and decomposers use organic molecules for their carbon sources. These organisms are called heterotrophs or organotrophs.

The autotrophs must fix the CO_2—that is, combine it with other atoms to make the organic molecules needed for cell growth. CO_2 fixation occurs in the Calvin-Benson cycle, where CO_2 is used to produce glyceraldehyde 3-phosphate, which can be used to produce glucose. Autotrophs use this glucose in the same glycolytic pathway used by heterotrophs.

The heterotrophs get their organic molecules from other organisms. These organic molecules are usually large carbohydrates, lipids, or proteins, which must be hydrolyzed into smaller molecules. The small molecules are then reassembled into the carbohydrates, lipids, and proteins that the cell needs.

Energy Sources

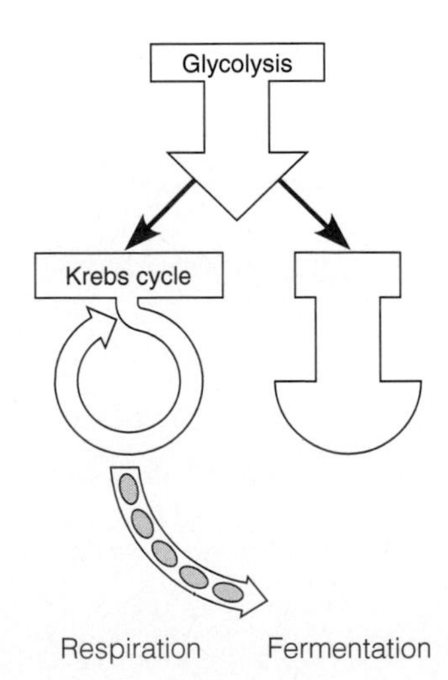

The figure to the right is an inset in many of the figures in Chapter 5 in your textbook. This figure provides an overview of the major energy-producing pathways in cells. Keep it in mind as you organize your new information on metabolism.

During the 1920s, a Dutch microbiologist, Albert Jan Kluyver, began looking for a simpler way to study metabolism. He made a list of the metabolic pathways that were known at that time. His list was very long, with specific chemical reactions carried out by different species. He then grouped the pathways into the following categories:

- Using organic molecules and combining them with oxygen

- Using organic molecules but not using oxygen
- Using inorganic compounds

Looking at his list, he saw the "unity in biochemistry." He concluded that all energy-yielding metabolism, whether heterotrophic or autotrophic or phototrophic or chemotrophic, were oxidation reactions. Oxidation is the removal of electrons from an atom or molecule. In biological systems, the electrons are often accompanied by a proton in a hydrogen atom. Look at Figure 5.26 (p. 141) in your textbook. This is a concept map showing the different nutritional types and what each uses for an energy source and a carbon source.

Chemotrophs oxidize chemicals to get energy. Most chemoheterotrophs oxidize glucose to produce ATP. Other carbohydrates such as starch, cellulose, and lactose are more available than glucose, so organisms have enzymes to hydrolyze these larger carbohydrates to produce glucose. Different species have different enzymes; consequently, they use different carbon sources and don't compete for resources; ultimately, all naturally occurring organic molecules are degraded. For example, many animal cells have amylases to hydrolyze starch into glucose and maltose. You are probably aware when your "blood sugar" (that's glucose) level is low and you have little energy. Eating a starch, such as bread or rice, will increase your blood glucose level as intestinal cells digest the starch. Some fungi and bacteria have cellulase to hydrolyze cellulose into glucose. The bottom line is that these larger carbohydrate molecules must be split to provide glucose.

Once inside a cell, glucose is oxidized in glycolysis. During glycolysis, energy is liberated to produce 2 molecules of ATP, and 2 molecules of NAD are reduced.

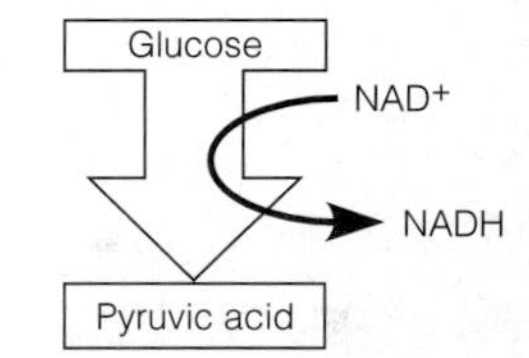

The NADH must be reoxidized to supply NAD^+ for glycolysis to continue. NADH can give electrons to an organic molecule in a process called fermentation. The pyruvic acid produced by glycolysis can be used to accept the electrons taken from the glucose. Lactic acid is produced when pyruvic acid accepts the electrons.

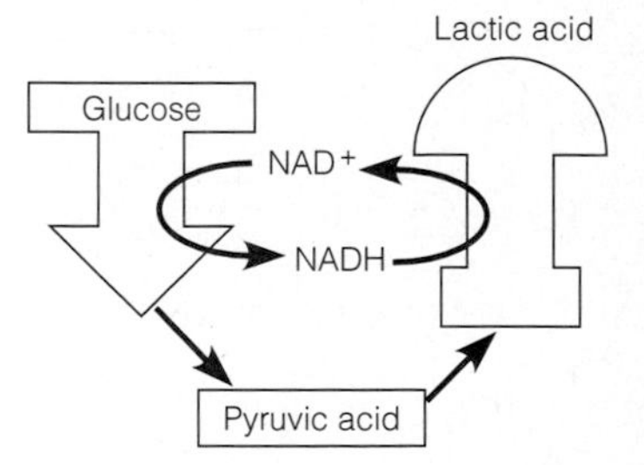

Bacteria and yeast can make a variety of fermentation end products by using other organic molecules as the final electron acceptor. Some of these products are listed in Table 5.17 (p. 133) of your textbook. Several of the fermentation products shown have been used in industrial processes. *Propionibacterium* is used to make Swiss cheese, and *Clostridium* is used to make butanol for synthetic rubber. You'll see that the end products of fermentation are often organic molecules such as ethyl alcohol and lactic acid. You know that ethyl alcohol still contains energy—you can see this when alcohol burns. Fermentation does not extract all of the available energy from the glucose molecule; its function is to oxidize the NADH produced during glycolysis. Remember that

Glycolysis

NAD^+ must be available for hydrolysis of another glucose molecule. In a laboratory, bacteria are often identified by the enzymes they have that (1) provide simple sugars for glycolysis and (2) produce fermentation products. For example, (1) *E. coli* can hydrolyze lactose to produce sugars for glycolysis, and *Salmonella* spp. cannot hydrolyze lactose; (2) *E. coli* produces lactic acid from fermentation, and *Enterobacter* produces acetoin.

Some organisms can get more energy from glucose by oxidizing pyruvic acid to CO_2 in the Krebs cycle. Note that this is the CO_2 you exhale. During the Krebs cycle, 1 ATP is produced by substrate-level phosphorylation, and more NADs are reduced. The primary purpose of the Krebs cycle is not to produce energy but to remove more electrons from pyruvic acid.

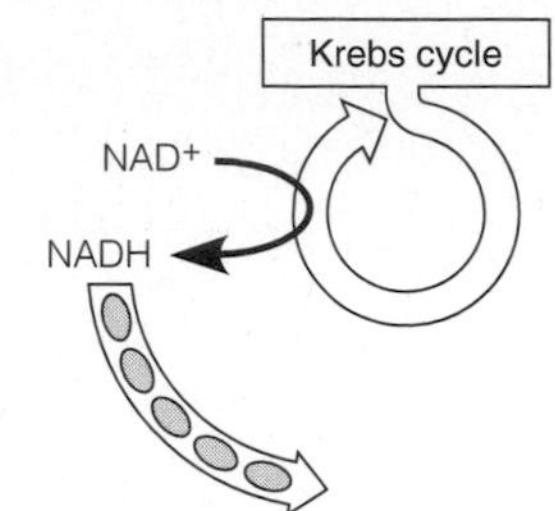

These organisms can get as many as 36 *more* ATPs from the glucose molecule because the electrons go to an electron transport chain. The electron transport chain, cellular respiration, is a series of oxidation-reduction reactions: electrons are passed from one carrier molecule (reduced) to an oxidized molecule. As the electrons are passed along the chain of carriers, energy is released that can be used to produce ATP.

An inorganic molecule usually serves as the final electron acceptor at the end of the electron transport chain. In aerobic respiration, molecular oxygen (O_2) is the final electron acceptor. Remember that the O_2 is not providing energy; its function is to dispose of the electrons. The O_2 picks up electrons and protons to form H_2O, one of the end products of respiration. Look at Figure 5.16 (p. 313) in your textbook to see a summary of aerobic energy production.

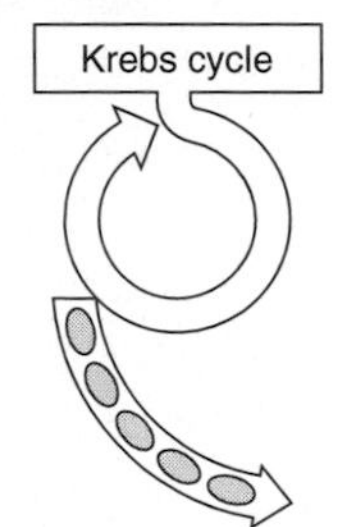

Some bacteria—such as those living in your intestines, the ocean sediments, and other anoxic environments—are able to combine electrons with inorganic ions or compounds other than O_2; this is called anaerobic respiration. These cells get more energy from glucose than they would if they used fermentation alone.

Some chemotrophs oxidize inorganic chemicals such as iron or sulfur to get electrons for their electron transport chains. Remember that these organisms will still need glucose or CO_2 for a carbon source and will use part or all of glycolysis and the Krebs cycle to make new cellular material.

S^{2-} S^0

Sunlight drives the electron transport chain of phototrophs. Chlorophyll emits electrons when it absorbs light. These electrons then go through an electron transport chain to produce the ATP needed for CO_2 fixation. This electron transport chain is not the same one used in respiration, but the principle is the same. Electrons are passed from a reduced carrier to an oxidized carrier, and energy from the transfers is used to generate ATP. The electrons ultimately return to chlorophyll in cyclic photophosphorylation. In noncyclic photophosphorylation, another electron donor is needed to return electrons to chlorophyll. Most phototrophs split H_2O to get electrons. A by-product of this is the production of O_2. A few bacteria can split H_2S to get electrons for chlorophyll.

$$2H_2O \longrightarrow O_2 + 4H^+ \text{ (to NAD) and } 4e^- \text{ (to chlorophyll)}$$

Sunlight provides the energy for phototrophs, but they need a carbon source. Photoautotrophs use CO_2 as their carbon source, and photoheterotrophs use an organic compound for carbon.

All of the enzymes required for these metabolic pathways are encoded in a cell's DNA, which is the subject of Unit 4, Reading the Code of Life.

Cell Growth

Most prokaryotic organisms grow by binary fission. As shown in Figure 6.11 (p. 169) in your textbook, a cell grows for a while, and its chromosome replicates. As the cell grows longer, the duplicate chromosomes separate, and the plasma membrane pinches in to form two offspring cells. Bacterial cells divide at their fastest rate with optimum temperature, nutrients, and oxygen. Generation times range from as fast as 10 minutes for *Vibrio natriegans* to several hours for *Mycobacterium tuberculosis*. Try the following problem to appreciate how fast bacteria grow.

Assume 1 *E. coli* O157:H7 cell was inoculated into the leftover beef stew you planned to have for lunch and that the bacterium started growing when you left for work at 7:00 A.M. If it had a generation time of 30 minutes, how many *E. coli* bacteria would be in the stew by noon? 1024; the ID_{50} is ~50 cells.

Bacteria grow quickly under optimum conditions. But optimum conditions are not often available outside of the laboratory. As the individuals of one species grow, they produce waste products. The accumulation of waste products with a concomitant decrease in nutrients leads to a stationary phase and, finally, a death phase for that population. Usually, not all of the members of that species will die; a few will survive and begin to grow when the right conditions recur. In the meantime, the waste products and dead cells provide nutrients for other species. This successive growth of different organisms recycles elements so they can be used again.

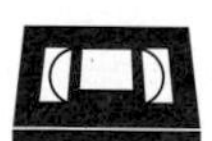

VIDEO: METABOLISM

Term

The following new term is introduced in this video:

chemolithotroph

Preview of Video Program

The importance of metabolism to life is introduced in the video program. You will see a variety of environments from natural deep-sea vents to manufactured wastewater treatment plants. Microbes are metabolizing in all of these places to produce energy for their growth. Some of the environments such as Mt. St. Helens and the sea vents are called extreme habitats.

Extreme habitats are too hot, acidic, or salty for most organisms. There are, however, prokaryotic organisms living in these environments. Most of these are the Archaea.

Applied microbiology is also introduced in this unit with visits to a wastewater treatment facility and a brewery. Applied microbiology is the use of microorganisms to produce food, industrial enzymes, and pharmaceuticals as well as the use of microorganisms in treating wastewater to remove pollutants.

Nancy Love describes the activated sludge process for sewage treatment. Settleable solids are removed from wastewater by primary treatment. In the secondary or biological process, bacteria metabolize dissolved organic compounds and produce smaller, inorganic compounds such as CO_2. The solids collected in primary treatment are moved to an anaerobic sludge digester, where bacteria ferment organic compounds and produce smaller organic compounds such as methane and ethyl alcohol.

Brewmaster Jamie Emerson describes another use of fermentation: to make beer. The fermentation process is quite similar to the metabolism occurring in an anaerobic sludge digester, except that the raw materials are carefully measured sugars and the microbe is a specific yeast that will produce ethyl alcohol and CO_2.

In the video, an animation shows the Submersible Vessel *Alvin* carrying scientists deep under the ocean to explore hydrothermal vents. Then Jim Leinfelder goes to a similar extreme environment on the Earth's surface at Mt. St. Helens. In both environments, chemolithotrophic (also called chemoautotrophic) bacteria use sulfur as their energy source. These bacteria oxidize the sulfur to get electrons for their electron transport chains. These bacteria change the sulfur so they can no longer use it, but now it is available for another organism to use:

$$S^{2-} \xrightarrow{\text{Species 1 using the S as an electron donor}} S^0 \xrightarrow{\text{Species 2 using the S as an electron acceptor}} S^{2-}$$

Recall from Unit 1 that each organism occupies a specific ecological niche depending on its carbon and energy needs. The point is made in the video that organisms change their environment as they grow and, thus, create niches for different organisms. These changes may mean that the organisms can't grow any longer—but now an organism that can use the new environment is able to grow. One such example occurs in an orchard when ripe fruit falls to the ground and is decomposed:

$$\text{Sucrose} \xrightarrow{\textit{Saccharomyces cerevisiae}} \text{Ethanol} \xrightarrow{\textit{Acetobacter}}$$
$$\text{Acetic acid} \xrightarrow{\textit{Desulfovibrio}} CO_2 \xrightarrow{\text{Plants}} \text{Sucrose}$$

Notice that plants have produced sucrose and the yeast will be able to grow again.

Jamie Emerson shows the standard culture technique used to grow bacteria and fungi in the laboratory. He sterilizes an inoculating loop by holding it in a flame. Using the sterile loop, he spreads an inoculum of microbial cells onto the surface of a culture medium, then incubates the plate to allow the bacteria or fungi to grow. Emerson incubates some plates aerobically and others in an anaerobic incubator. After incubation, many colonies are visible on the culture medium.

Draw the pattern of yeast colonies on Emerson's plate and circle one colony. His method of inoculation (or streaking) is slightly different from that shown in Figure 6.10 (p. 168) in your textbook, but the principle is the same. The medium is inoculated and, theoretically, each cell on the medium gives rise to one colony. As Emerson says, "We can take a single cell here and begin growing it." Emerson is using a differential medium so he can differentiate between the organism he wants and undesirable microbes.

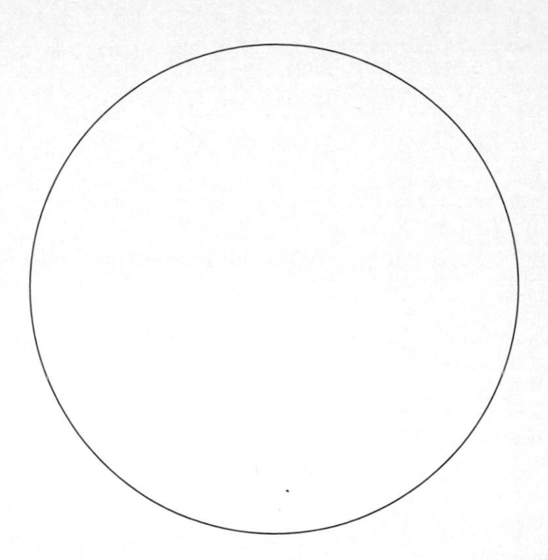

Video Questions

1. This video begins with a microscopic view of a growing colony of bacteria. What shape are the bacteria? Rods.
2. Write definitions of the following terms as you watch the video: (As shown on slates in the video.)
   a. Anabolism. The process of using the energy and simple building blocks to make new macromolecules for the cell.

   b. Catabolism. The process of breaking down macromolecules into their smaller component parts.

   c. Chemolithotrophy. Converting inorganic chemical energy such as hydrogen gas or sulfur into ATP.

   d. Fermentation. The conversion of organic chemical energy, such as organic waste or sugars, into ATP; does not use an electron transport chain; the final electron acceptor is usually an organic molecule.

   e. Photosynthesis. The conversion of light energy into ATP.

   f. Respiration. The conversion of organic chemical energy, such as organic waste or sugars, into ATP; uses an electron transport chain; the final electron acceptor is usually an inorganic atom, ion, or molecule.

3. The source of energy for primary producers such as plants and photosynthetic bacteria is light.

   What are the primary producers in a deep-sea food chain? Chemolithotrophic bacteria.

   What do these organisms use for an energy source? Oxidation of inorganic chemicals.

4. Why do yeast make beer? To get energy for their growth.

5. As you watch the video, fill in this graph to show a typical bacterial growth curve. Label the lag, logarithmic, stationary, and death phases. What is happening in each phase? Lag: Cells are synthesizing DNA and enzymes. Log: Maximum growth rate under conditions provided. Stationary: The number of cells dying equals the number of cells dividing; many cells have stopped dividing. Death: The number of cells dying is greater than the number that are dividing.

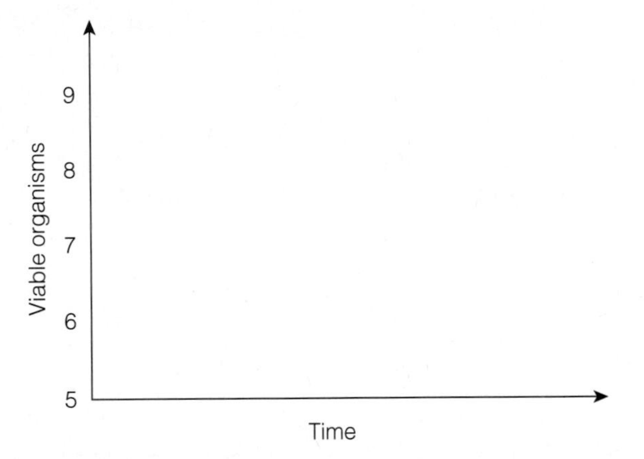

6. What is happening in each of these places in a sewage treatment plant? Which steps require air (O_2)? Where does the effluent go at ❶? Where does the waste go at ❷? Sedimentation: Separation of solids from water. Activated sludge: Aerobic digestion of dissolved organic compounds. Anaerobic digester: Microbial digestion of solid organic compounds. ❶ Water goes to receiving water, such as a river or ocean, or onto land for irrigation. ❷ Solids go to a dump or to agricultural land.

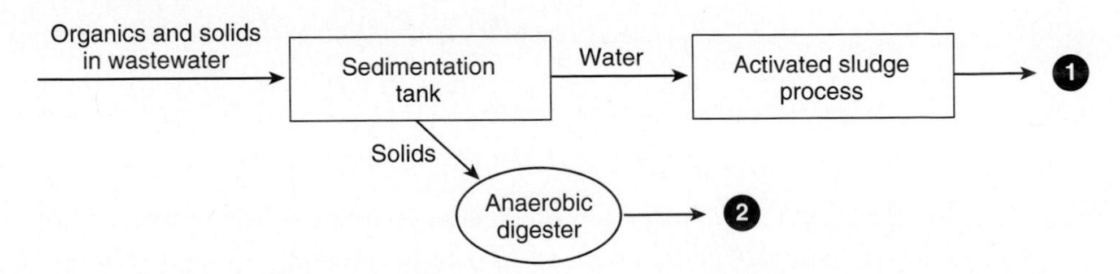

7. Why does this video include a study of sewage treatment and beer making? Both industries use the same metabolic process, fermentation.

EXERCISES

Concept Map 3.1

This map shows terms related to cellular metabolism.

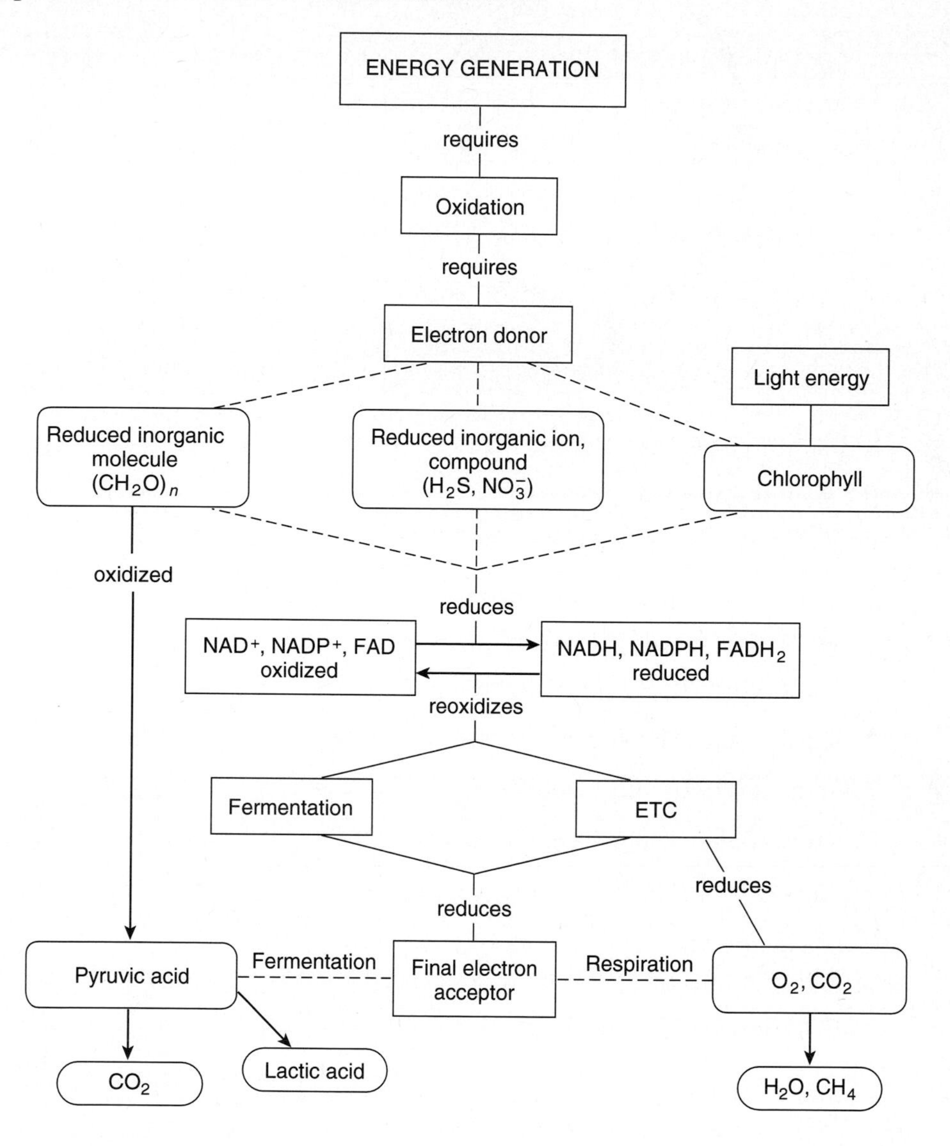

1. Does this map show catabolic or anabolic pathways? ____________________
2. All energy generation requires ____________________.
3. Why is it necessary to have fermentation or the electron transport chain after glycolysis?

4. Mark the glycolytic pathway on the concept map.
5. Mark the cellular respiration pathway on the concept map. Mark the place(s) where aerobic and anaerobic respiration differ.

Figure 3.1

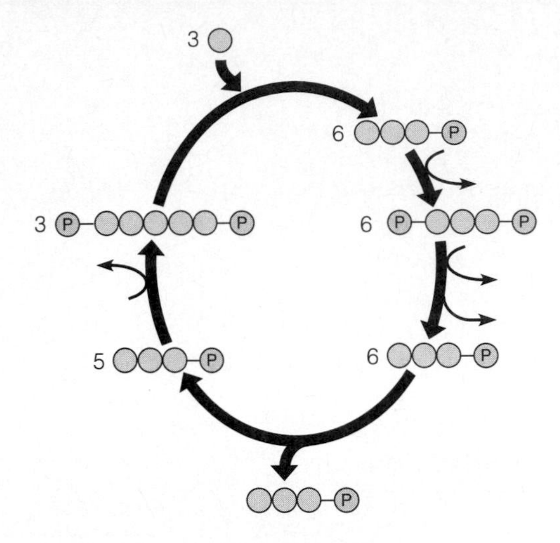

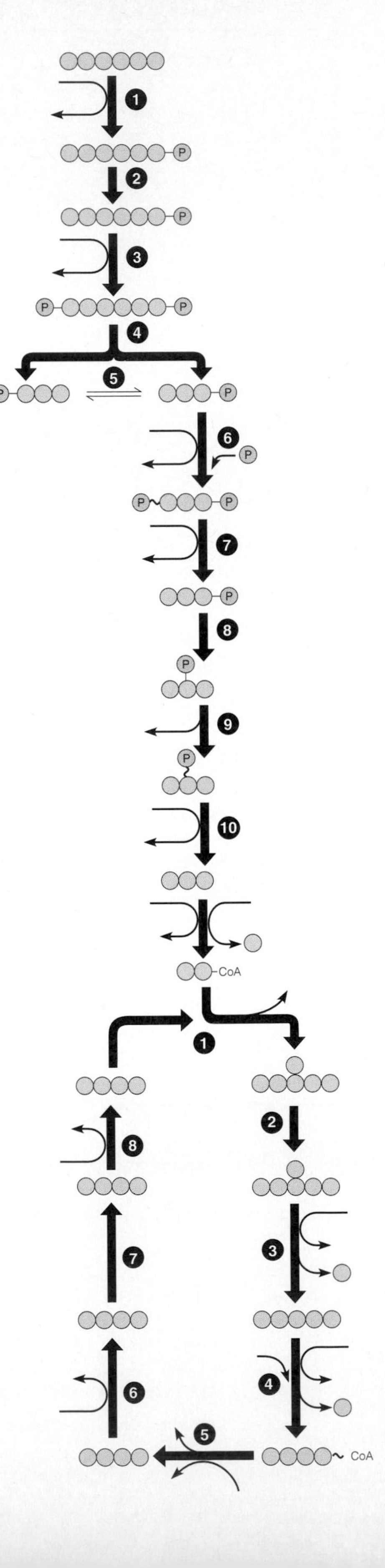

1. Identify the Calvin-Benson cycle.
2. Identify glycolysis.
3. Identify the Krebs cycle.
4. Show how the Calvin-Benson cycle can supply glucose for glycolysis.
5. Show where ATP is used in these pathways.
6. Show where ATP is made in these pathways.
7. Show where lipid catabolism occurs.
8. Show where protein catabolism occurs.
9. Show where lipid synthesis occurs.
10. If a heterotrophic cell completely oxidizes 6×10^6 molecules of glucose, how many molecules of CO_2 does that cell make?

11. Assume you are growing a photosynthetic cell in a test tube. The cell produces 6 molecules of glyceraldehyde 3-phosphate and uses 3×10^6 molecules of glucose in the Krebs cycle while you are studying it. What is the net change in CO_2 in the test tube?

Figure 3.2

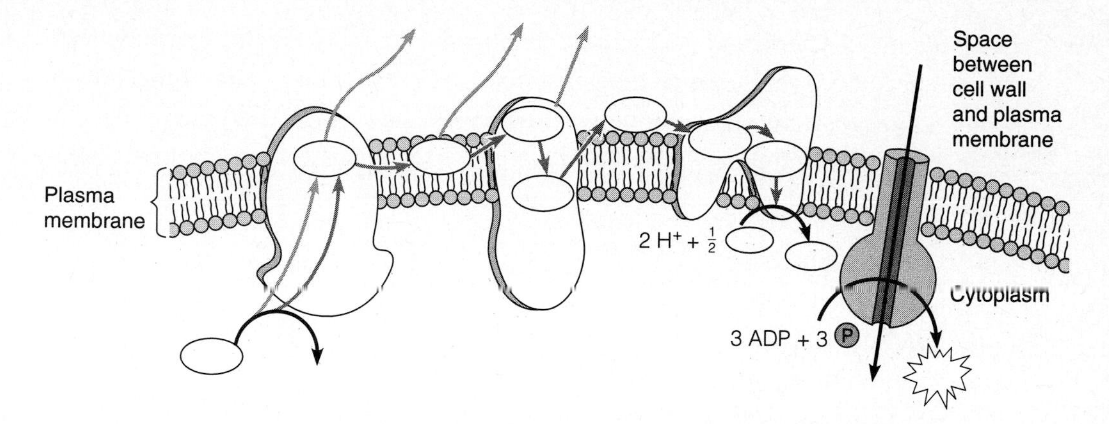

Identify the following:

1. The electron pathway.
2. The positively charged side of the membrane.
3. The acidic side of the membrane. (Read about acids on p. 38 of your textbook.)
4. The sites where protons are pumped across the membrane.
5. The site were ATP is made.
6. The proton pathway.
7. The protons and electrons that combine with oxygen.
8. How this pathway could be modified to show anaerobic respiration.

Definitions

Match the following statements to words from Key Terms and Concepts:

1. Proteins that speed up chemical reactions. ____________________
2. A cell that uses CO_2 at the end of its electron transport chain. ____________________
3. Cells that use glucose as a carbon and energy source. ____________________
4. Cells that use CO_2 as a carbon source and sulfur as an energy source. ____________________
5. A coenzyme used in glycolysis. ____________________
6. The removal of CO_2 from an amino acid. ____________________
7. A product of the light reactions of photosynthesis that is needed by the dark reactions. ____________________
8. The purpose of the dark reactions of photosynthesis. ____________________
9. How ATP is made in glycolysis. ____________________
10. Excess fluoride ions can kill cells because they act like these. ____________________

Matching Questions

To identify bacteria and yeasts, microbiologists indirectly test for genes by testing for the presence of enzymes, which are gene products. In the video, Jamie Emerson could differentiate *Saccharomyces* yeast from unwanted organisms by the metabolic products they produced on a culture medium. Use the following processes to answer the questions below:

a. Calvin-Benson cycle
b. Deamination
c. Decarboxylation
d. Electron acceptor
e. Electron transport chain; aerobic respiration
f. Electron transport chain; anaerobic respiration
g. Fermentation
h. Glycolysis
i. Krebs cycle

1. *Salmonella* can grow in the presence of cyanide. Cyanide inhibits cytochrome oxidase. What pathway(s) are they using? ____
2. *E. coli* can reduce nitrate ion (N^{+5}) to nitrite ion (N^{+3}). What pathway is it using? ____
3. *Pseudomonas* can use glucose aerobically but not anaerobically. What pathway(s) are they using? ____
4. *Staphylococcus aureus* produces acid from mannitol. By what pathways? ____
5. *Streptococcus* lack cytochromes. How do they produce energy? ____
6. *Citrobacter* produce acid from lactose. By what pathways? ____
7. *Hemophilus* need NAD^+ provided in its growth medium. For what do they need NAD^+? ___
8. *Bacillus* can hydrolyze starch. What do they do next? ____
9. *Proteus vulgaris* removes —NH_2 from the amino acid ornithine. What is this called? ___
10. *Streptococcus pyogenes* produces acid from sorbitol. By what pathways? ____
11. The unicellular eukaryote *Euglena* is a photoautotroph when it grows in the light but a chemoheterotroph when it grows in the dark. What pathway(s) is it using in the light that it is not using in the dark? ____

Comparison Questions

Assume you are growing two yeast cultures in a glucose broth. Culture A is incubated aerobically at 30°C for 48 hours, and culture B is incubated anaerobically at 30°C for 48 hours. Use the following choices to answer the questions below:

A. Culture A

B. Culture B

C. Both cultures A and B

D. Neither culture A nor B

1. Which culture is using oxidative phosphorylation? ____
2. Which culture is using substrate-level phosphorylation? ____
3. Which culture produces the most ATP? ____
4. Which culture is using NAD^+? ____
5. Which culture is using photophosphorylation? ____
6. Which culture produced the most cells? ____
7. Which culture is using glycolysis? ____
8. Which culture is using an electron transport chain? ____
9. Which culture is using the Calvin-Benson cycle? ____
10. Which culture is using anabolic pathways? ____

Fill-In Questions

Use the following choices to complete the statements below. Choices may be used once, more than once, or not at all.

11	denature	NAD^+
89	electron transport chain	O_2
95	glucose	oxidation
activate	glycolysis	pyruvic acid
Calvin-Benson cycle	H_2O	reduction
CO_2	Krebs cycle	
deaminate	lactic acid	

1. In glycolysis, ____________________ is oxidized to pyruvic acid.
2. Acetyl is oxidized in ____________________ to produce ____________________.
3. ____________________ % of the ATPs made from the complete oxidation of glucose by a prokaryotic cell is made by oxidative phosphorylation.
4. The removal of electrons from an atom is ____________________.
5. The final electron acceptor in lactic acid fermentation is ____________________.

6. The high temperature of a deep-sea vent or low pH of a hot spring will ______________________ enzymes of most organisms.
7. Cellular respiration requires ______________________.
8. In order to be used in the Krebs cycle, a cell must ______________________ an amino acid.
9. O_2 is produced by cyanobacteria and plants from the splitting of ______________________.
10. Autotrophs use ______________________ as a carbon source.

Short-Answer Questions

1. Define anabolism and catabolism. Why is ATP called an "intermediate of metabolism"?

2. What is the purpose of an energy source? Why isn't ATP the energy source?

3. What is the purpose of a carbon source?

4. Yeast are added to a sugar solution and incubated anaerobically to make wine. What would happen if the mixture was incubated aerobically?

5. What is the purpose of fermentation? Of the Krebs cycle?

6. Compare and contrast oxidative phosphorylation and photophosphorylation.

7. Compare and contrast *E. coli* using glucose as an energy source and *Thiobacillus* using sulfur as an energy source. Both bacteria are growing aerobically.

8. Compare and contrast fermentation and anaerobic respiration.

9. What effect will it have on energy production if a cell removes α-ketoglutaric acid from glycolysis to make the amino acid glutamic acid?

10. Differentiate between *Beggiatoa* bacteria, which use sulfur (as sulfide ion, S^{2-}) as an electron donor, and *Thermofilum* archaea, which use elemental sulfur (S^0) as an electron acceptor.

Hypothesis Testing

Students can post their evidence on an electronic bulletin board to arrive at a class conclusion.

Hypothesis: The first organisms on Earth were chemoheterotrophs.

Data/Observations: What evidence supports this hypothesis? Living organisms are made of organic compounds. Both heterotrophs and autotrophs use glycolysis. Autotrophic metabolism is more complex than heterotrophic metabolism.

Conclusions: From these data, do you accept or reject the hypothesis? Briefly explain.

Study Questions

***Microbiology: An Introduction*, Sixth Edition**

Pages	Review Questions	Multiple Choice Questions	Critical Thinking Questions	Clinical Applications Questions
51–153	All	All	All	All
740	11	3		
741		6–8		
759–760	5–7, 10–11	7, 9, 10	1, 2, 3	
54–55	All	All	All	All

CD Activities

1. Do the Chapter 5 quiz.
2. Use the Interactive Unit on Metabolism to get hands-on practice with metabolism. The Metabolism unit includes activities on oxidization-reduction reactions as well as glycolysis, fermentation, the Krebs cycle, and respiration. There are also experiments you can do to learn more about enzymes.

3. Use the Interactive Unit on Growth to get practice drawing growth curves and analyzing bacterial growth.
4. Use the Chemistry Unit to review chemical principles.

Web Activities

1. Do the Chapter 5 quiz.
2. Read "Unique Ecological Niche of *Desulfovibrio*," and answer the following questions:
   a. What does *Desulfovibrio* use for its energy source? Its carbon source? Small organic molecules such as lactic acid.

   b. Is it a chemoheterotroph, chemoautotroph, photoheterotroph, or photoautotroph? Chemoheterotroph.

   c. Compare and contrast your cellular metabolism with that of *Desulfovibrio*. Both are chemoheterotrophs. Humans use large organic molecules for both their carbon and energy sources, and use aerobic respiration. *Desulfovibrio* uses sulfate as its electron acceptor in anaerobic respiration.

Lab Activity

Prepare one or more of the fermented foods described in Appendix B.

ANSWERS

Concept Map 3.1

1. Catabolic pathways
2. Oxidation
3. Fermentation and the ETC serve to reoxidize NADH.
4. The line on the left showing a $(CH_2O)_n \rightarrow$ pyruvic acid
5. Respiration is the ETC. Molecular oxygen (O_2) is the final electron acceptor in aerobic respiration; in anaerobic respiration, a different molecule, compound, or ion will serve as the final electron acceptor at the end of the ETC.

Figure 3.1

Questions 1–6 can be answered using Figures 5.11, 5.12, and 5.24 in your textbook.

7. See Figure 5.20.
8. See Figure 5.21.
9. See Figure 5.28.
10. 36×10^6 molecules of CO_2
11. 0 molecules of CO_2

Figure 3.2

Use Figure 5.15 in your textbook to answer these questions.

The following answers are not written in the figure and require analytic thinking:

2. The outside, with the accumulation of H^+.
3. The outside, with the accumulation of H^+.
8. Replace the O_2 with a different inorganic compound, such as sulfate ion (SO_4^{3-}).

Definitions

1. enzymes
2. anaerobe
3. chemoheterotrophs
4. chemoautotrophs
5. NAD^+
6. decarboxylation
7. ATP
8. carbon fixation
9. substrate-level phosphorylation
10. noncompetitive inhibitors

Matching Questions

1. g, h
2. f
3. e, i, h
4. g, h
5. h
6. g, h
7. d
8. h
9. b
10. g, h
11. a

Comparison Questions

1. A
2. C
3. A
4. C
5. D
6. A
7. C

8. A
9. D
10. C

Fill-In Questions

1. glucose
2. Krebs cycle, CO_2
3. 89
4. oxidation
5. pyruvic acid
6. denature
7. electron transport chain
8. deaminate
9. H_2O
10. CO_2

Short-Answer Questions

1. Anabolism uses energy (ATP); catabolism produces energy (ATP). The two processes are linked by ATP.
2. The energy source is used to generate ATP. ATP is not available in the environment. A cell converts the energy in its source to chemical energy in ATP.
3. The carbon source is used to make new cellular material.
4. Anaerobically, the yeast ferment the sugars to make ethyl alcohol and CO_2. Aerobically, the yeast will switch from fermentation to aerobic respiration. They will produce CO_2 and H_2O instead of ethyl alcohol and CO_2.
5. Fermentation reoxidizes NADH. The Krebs cycle produces NADH (electrons) for the electron transport chain.
6. *Compare:* Both processes use an electron transport chain and chemiosmosis to produce ATP. *Contrast:* The source of energy in oxidative phosphorylation is a chemical donor—e.g., glucose in heterotrophs. In photophosphorylation, the energy from light liberates electrons from chlorophyll.
7. *Compare:* Both organisms are using an electron transport chain. *Contrast: E. coli* is transferring electrons from glucose to NAD^+ in glycolysis and the Krebs cycle. *Thiobacillus* are transferring electrons directly from the sulfur atoms to NAD^+.
8. *Compare:* Both processes reoxidize NADH anaerobically. *Contrast:* Fermentation uses an organic molecule as the final electron acceptor and does not produce energy; no electron transport chain is involved. It follows the energy-producing glycolysis. Anaerobic respiration usually uses an inorganic molecule as the final electron acceptor and does produce energy via an electron transport chain.
9. The cell loses one turn of the Krebs cycle and the potential energy from that cycle.
   S^{2-} *Beggiatoa* S^0 *Thermofilum* S^{2-}
10. *Beggiatoa* are transferring electrons from sulfide to NADH. These electrons will be used to generate energy in the electron transport chain. *Thermofilum* are using sulfur at the end of the electron transport chain, anaerobically. *Thermofilum* will use glucose as an energy source (electron donor).

Unit 4

READING THE CODE OF LIFE

There are, however, other possibilities, notably those connected with naturally occurring substances. Little, however, has been done to purify or to determine the properties of any of these substances.
—Ernst Chain, 1940

Learning Objectives

	1.	Define genetics, chromosome, gene, genetic code, genotype, and phenotype.	p. 207
★	2.	Describe how DNA serves as genetic information.	p. 207
	3.	Describe the process of DNA replication.	p. 209
★	4.	Describe protein synthesis, including transcription, RNA processing, and translation.	p. 211
★	5.	Explain the regulation of gene expression in bacteria by induction, repression, and catabolic repression.	p. 215
	6.	Classify mutations by type, and describe how mutations are prevented or repaired.	p. 221
	7.	Define mutagen.	p. 221
	8.	Outline the methods of direct and indirect selection of mutants.	p. 226
	9.	Identify the purpose of and outline the procedure for the Ames test.	p. 227
★	10.	Discuss how genetic mutation and recombination provide material for natural selection to act on.	p. 237

Reading

Chapter 8 (pp. 207–229, 237) and pp. 547, 550–551, and 699. (Genetic Transfer and Recombination, pp. 229–236, are in Unit 5.)

Suggested Labs from Johnson and Case, *Laboratory Experiments in Microbiology,* Fifth Edition

Exercise 27: Isolation of Bacterial Mutants
Exercise 31: Ames Test for Detecting Possible Chemical Carcinogens

Key Terms and Concepts

DNA and Chromosomes
base pairs p. 208
chromosomes p. 207
DNA ligase p. 210
genes p. 207
genetic code p. 208
genetics p. 207
genotype p. 208
lagging strand p. 210

leading strand p. 210
phenotype p. 208
replication fork p. 210
RNA primer p. 210
semiconservative replication p. 210

RNA and Protein Synthesis

codons p. 213
corepressor p. 219
cyclic AMP p. 220
degeneracy p. 213
enzyme induction p. 219
inducer p. 217
induction p. 217
messenger RNA p. 212
nonsense codons p. 214
operator p. 219
operon p. 219
promoter p. 213
repression p. 217
repressors p. 217
RNA polymerase p. 213
sense codons p. 213
structural genes p. 219
terminator p. 213
transcription p. 212
transfer RNA p. 214
translation p. 213

Mutation

Ames test p. 228
auxotroph p. 227
base analog p. 224
base substitution p. 221
carcinogens p. 227
catabolic repression p. 221
excision repair p. 225
frameshift mutations p. 223
light-repair enzymes p. 225
missense mutation p. 222
mutagens p. 223
mutation p. 221
mutation rate p. 226
negative (indirect) selection p. 226
nonsense mutation p. 222
positive (direct) selection p. 226
replica plating p. 226
spontaneous mutations p. 223

INTRODUCTION

Be sure to read the Study Outline for Chapter 8, pp. 237–239.

The genome of all cells is composed of DNA, and all DNA is made of four repeating nucleotides: adenine (A), thymine (T), cytosine (C), and guanine (G). Think of the bases as letters in an alphabet: English has 26 letters, Spanish has 28 letters, Hawaiian has 12 letters, and the genetic alphabet has 4 letters. The letters are arranged to make a code or words; and the words are strung together in sentences to express an idea. In genetics, the letters and their codes are arranged into genes, which are expressed to make products to produce the cell's phenotype. Genes differ from each other because the letters are used in different sequences.

Genes are not physical structures you can see. They are regions of DNA, "seen" by RNA polymerase, that code for a functional product: a protein, or RNA. Some gene products are structural, such as a pigment or flagella. Other gene products have more subtle activity, such as the catabolic activator protein that regulates expression of other genes.

To express a gene, a copy must be made by a process called transcription. This is similar to a student copying or transcribing what the instructor has written on the chalkboard (the DNA). The transcript is in the form of a molecule called RNA. Usually, this RNA code is changed into another "language" or translated, resulting in a protein. This drawing is used as an inset throughout Chapter 8 to orient the process being studied. The circular DNA represents a prokaryotic chromosome. Although eukaryotic DNA is arranged in linear chromosomes, the processes of transcription and translation are the same in both types of cells. Label the chromosome, mRNA, and protein. Show where transcription and translation are occurring.

Protein synthesis is ongoing in a cell, but not all genes are transcribed all of the time. Transcription of some genes is stopped by a repressor protein. Repressor proteins are made by regulatory genes. Transcription of other genes can be started or induced by the presence of an inducer molecule.

The genetic code, depicted in Figure 8.7 (p. 215) in your textbook, shows that the genetic "words" consist of three nucleotides, which code for amino acids. Although there is degeneracy in the code, which protects a cell against the effects of mutations, some mutations may change an amino acid in a protein.

A common misconception is that mutations (and evolution) are planned. There is no grand scheme. Rather, organisms that happen to be fit to survive in a particular environment at a particular time will survive and reproduce until they get an unfavorable mutation or the environment changes. Imagine that a bacterial cell is suddenly exposed to an antibiotic. The antibiotic may kill the cell in minutes—before the cell can make, transcribe, and translate a new gene. In 1943, Salvador Luria and Max Delbrück demonstrated that mutations occur randomly, and if they happen to give the cell an advantage sometime, that's nice, but it isn't necessary that the mutation have any advantage or that a new gene product ever be used.

Antibiotic resistance in bacteria is a major problem requiring the discovery of new antibiotics. In this unit's opening quote, one of the codiscoverers of penicillin suggested that there are more antibiotics to be found. Current research on new antibiotics involves doing just that. In the video, you will see two projects: one involving woodland soil and another using radiation-contaminated soil. Researchers in both projects are screening bacteria for the production of antibiotics, an approach first developed by Selman Waksman when he discovered the bacterium that produces streptomycin in 1944.

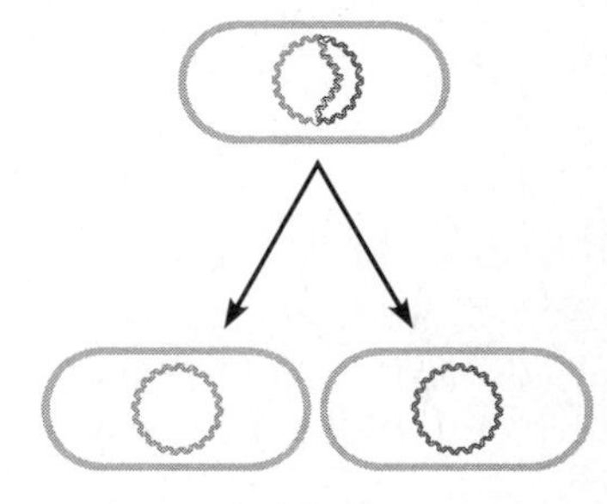

This figure from your textbook shows a cell replicating. Before dividing, a cell must have two complete copies of all of its genes to distribute to its daughter cells. Label the replication fork. The circular chromosome indicates that this is a prokaryotic cell. Chromosome replication is the same in eukaryotic cells; however, eukaryotic cells have an elaborate process, called mitosis, to ensure that one of each different chromosome gets to each new cell.

Methods of genetic recombination and application of these natural processes in biotechnology are the subjects of Unit 5, Genetic Transfer.

VIDEO: READING THE CODE OF LIFE

Terms

The following new terms are introduced in this video:

antibiotic
antibiotic resistance

Preview of Video Program

This video program begins with an introduction to DNA and the genetic code. DNA carries the genes that code for a cell's proteins and RNA. Protein synthesis is described.

DNA may undergo changes called spontaneous mutations as a result of mistakes made during DNA replication. Mutations are rare events that are usually corrected by repair enzymes. An uncorrected mutation can be lethal if it results in loss of function or beneficial if it results in a useful protein. Mutations that led to antibiotic resistance in bacteria are beneficial to these bacteria, as they can now grow in environments without competition from other bacteria. Bacteria will pass these mutated genes to their offspring and can pass them to other cells. Julian Davies lists some reasons for the prevalence of antibiotic resistance. He talks about his project screening soil bacteria for new antibiotic-producing bacteria. Davies is looking for naturally occuring bacteria that have not yet been studied. And Jenny Hunter-Cevera and Yuri Gleba are looking for mutant bacteria in irradiated soil near the Chernobyl nuclear reactor. The rate of mutations is accelerated by radiation, so these bacteria may be expressing many new products, including antibiotics.

Researchers at the Kiev Institute of Microbiology are using traditional methods for culturing bacteria and assaying for antibiotics, but the location—radiation-contaminated soil—is new. Davies and his colleagues are incorporating new molecular biology techniques into their screening of common woodland soil. Davis is investigating bacteria that cannot be grown in the lab, so he puts their genes in cells that can be grown in the lab. Davies's method uses the following steps:

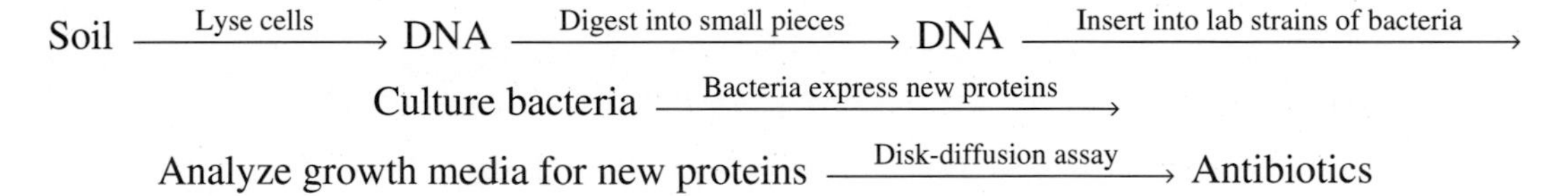

The process by which DNA can be inserted into cells will be discussed in Unit 5.

Constitutively expressed genes, such as glycolytic enzymes, are made all the time in a cell. To conserve energy and resources, however, many genes are transcribed only when the protein is needed. These genes may be repressible or inducible. Transcription of repressible genes is stopped by a repressor protein; transcription of inducible genes is turned on when the gene products are needed. Consider *E. coli* growing in the human intestine, dependent on the eating habits of its human host. If the host takes in a tryptophan-rich meal, the bacteria do not have to make tryptophan. On the other hand, if the host has a tryptophan-deficient diet, the bacteria will need to make their own tryptophan by transcribing and translating the genes for tryptophan synthesis.

Video Questions

1. Consider the phrase "And in 50 years' time, microbes have learned how to live with them [antibiotics]." Discuss whether the microbes have learned or have been selected. Students can discuss anthropomorphism as well as the inability of bacteria to learn.

2. List the factors that contribute to the selection of antibiotic-resistant bacteria. Overuse and misuse of antibiotics are cited by Davies. Also see p. 551 of the textbook.

3. Why do we see mutant bacteria more often than mutant animals? Bacteria reproduce faster than animals, so we see many generations of bacteria in a short time. Bacteria are haploid, so a mutation is more likely to be expressed.

4. Is all DNA in a cell active all the time? No.

5. Jim Leinfelder refers to the thermostat in the house. How is feedback repression like the thermostat? The product (heat) acts on the repressor (thermostat) to turn off expression of the gene (heat from the furnace).

EXERCISES

Concept Map 4.1

This map shows terms related to microbial genetics.

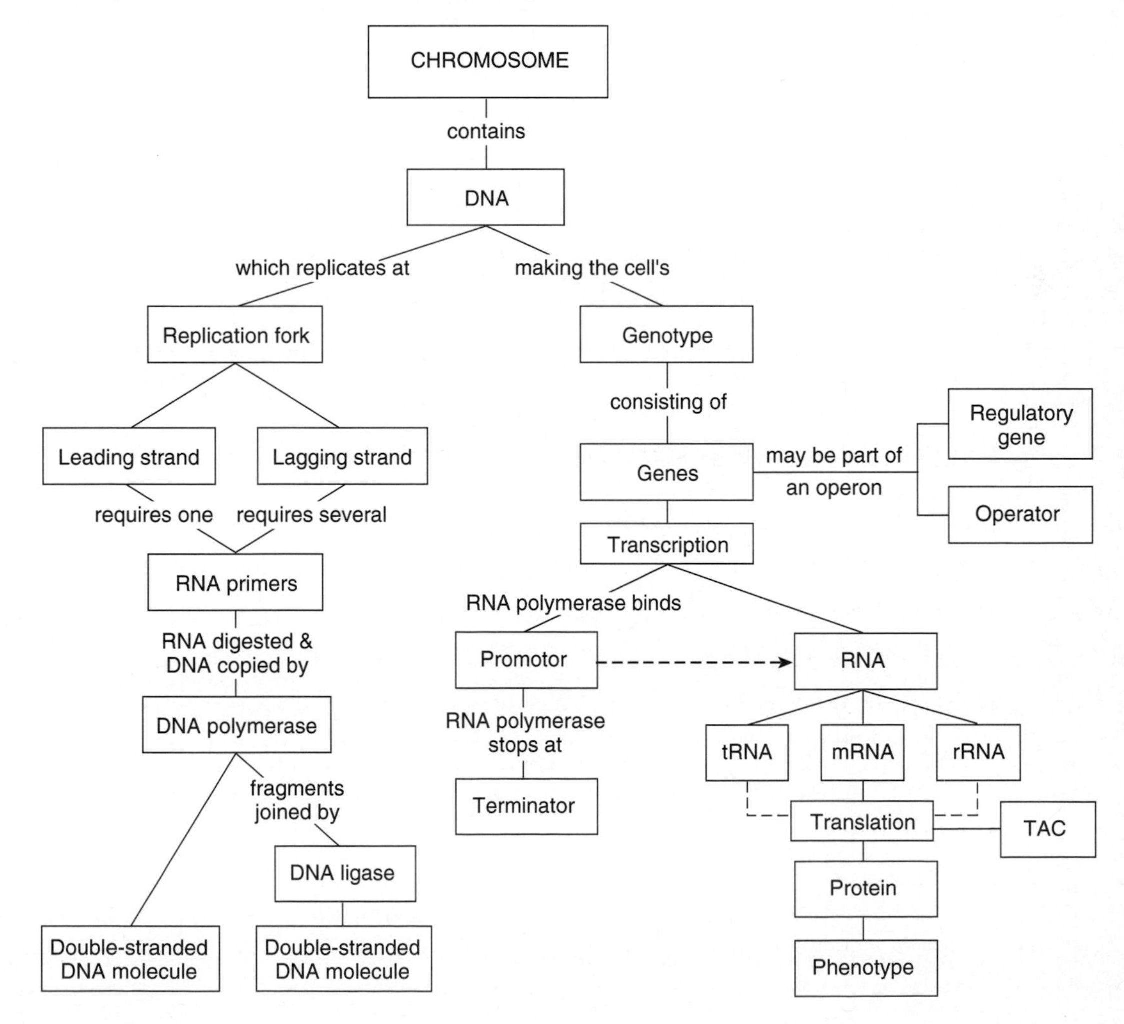

1. What are the products of transcription?
2. Place these two terms in the concept map. codon | stop codon
3. How does synthesis of the leading and lagging strands differ?

4. What happens at the replication fork?

Definitions

Match the following statements to words from Key Terms and Concepts:

1. DNA strand that is made continuously. ______________________
2. The characteristics of an organism resulting from expressed genes. ________________
3. A small molecule used to start DNA synthesis. ______________________
4. The enzyme that makes RNA. ______________________
5. Codons are read from this molecule. ______________________
6. DNA repair enzymes activated by visible light. ______________________
7. Insertion of nucleotides in a molecule of DNA. ______________________
8. The molecule that binds to the catabolic activator protein. ______________________
9. Replacing one nucleotide in DNA with another. ______________________
10. A mutant cell that requires a nutrient not required by the parent cell. ______________

Practice with Base-Pairing and Mutations

Cut out the nucleotide bases on p. 59 of this guide. This nucleotide pool is available for DNA and RNA synthesis.

1. Use these nucleotides to replicate the DNA shown on p. 57 of this guide. Remember that DNA can be synthesized only in the $5' \rightarrow 3'$ direction.
2. Use the original strand of DNA as a template for transcription. RNA polymerase can only go in one direction.
3. Translate the mRNA molecule you made into a peptide using Figure 8.7 (p. 215) in your textbook.
4. Repeat step 2 but replace one of the thymine nucleotides with AZT. Repeat again with 5-bromouracil replacing a thymine nucleotide. Finally, repeat step 2 with acyclovir replacing one of the guanine nucleotides.
5. What happens when AZT is present? For what is AZT used? (*Hint:* Use the index in your textbook.)

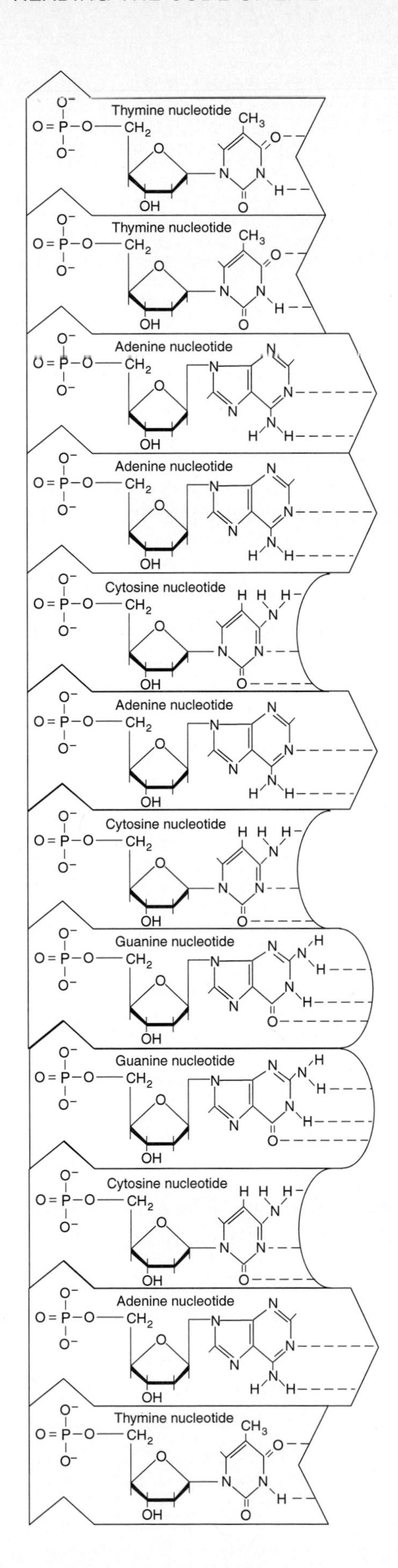
Thymine nucleotide
Thymine nucleotide
Adenine nucleotide
Adenine nucleotide
Cytosine nucleotide
Adenine nucleotide
Cytosine nucleotide
Guanine nucleotide
Guanine nucleotide
Cytosine nucleotide
Adenine nucleotide
Thymine nucleotide

6. What happens when acyclovir is present? For what is acyclovir used? (*Hint:* Use the index in your textbook.)

7. Why is 5-bromouracil a carcinogen?

Practice with Base-Pairing and Translation

A sense strand of DNA is shown below. The letter N represents the presence of a nucleotide base that is not needed for this question.

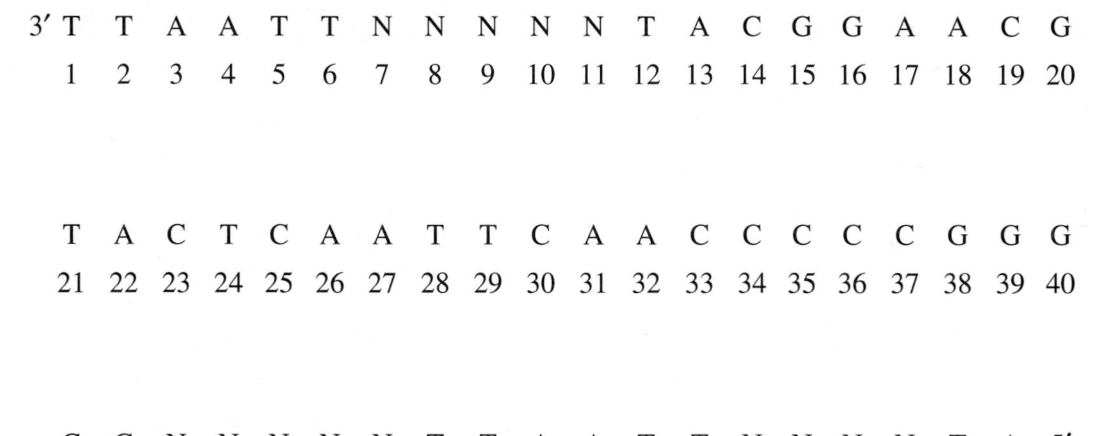

1. Write the complementary strand of DNA.
2. Using the strand given above as the sense strand, find the promoter to bind RNA polymerase. The enzyme will start transcription several bases downstream from the promoter at the start codon.
3. Start transcribing the DNA until you reach the terminator, which will stop transcription.
4. Use the genetic code in Figure 8.7 (p. 215) of your textbook to translate your mRNA into a peptide.
5. What would happen if a G were substituted for base 17?

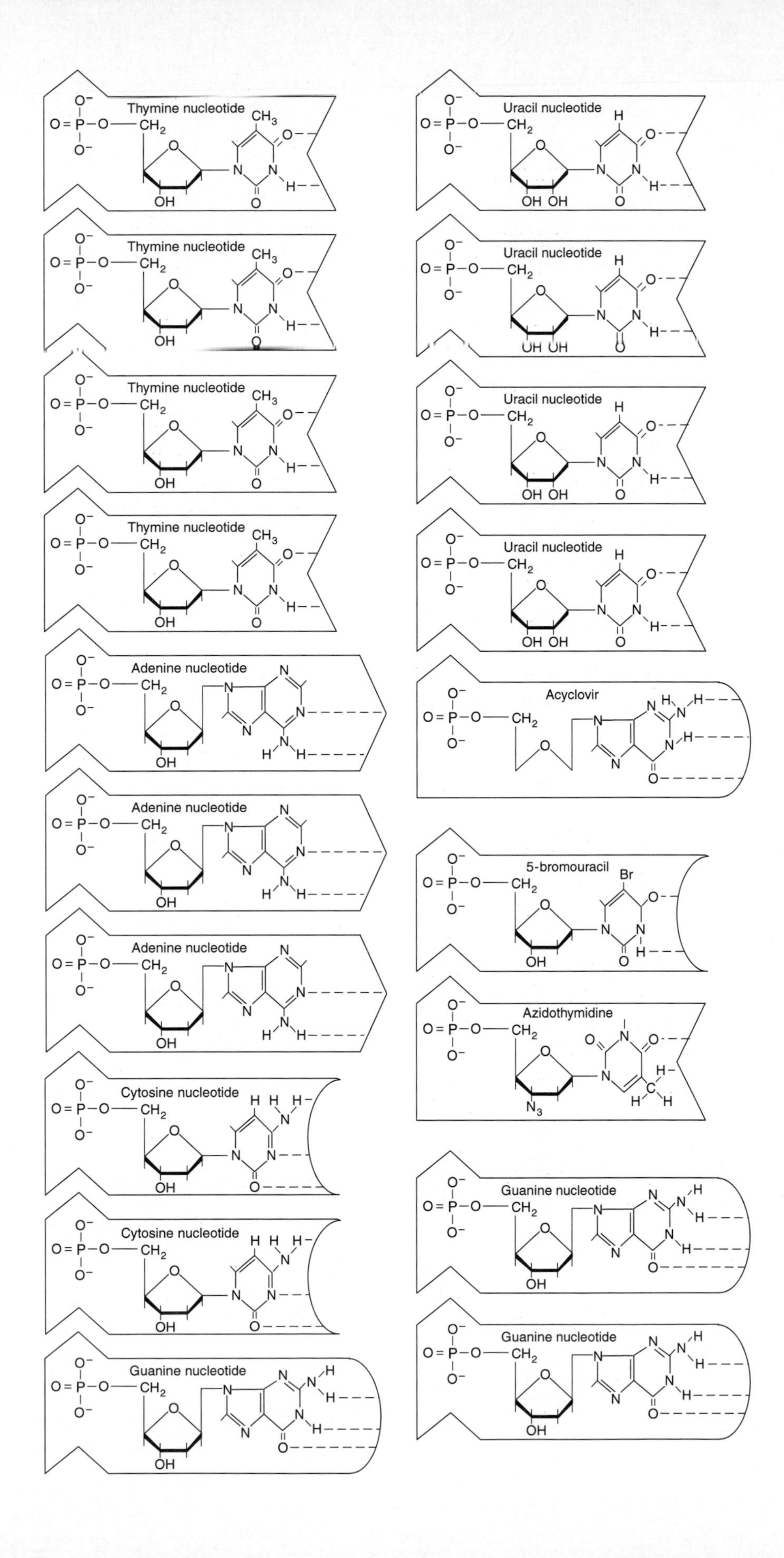
Thymine nucleotide
Thymine nucleotide
Thymine nucleotide
Thymine nucleotide
Adenine nucleotide
Adenine nucleotide
Adenine nucleotide
Cytosine nucleotide
Cytosine nucleotide
Guanine nucleotide
Uracil nucleotide
Uracil nucleotide
Uracil nucleotide
Uracil nucleotide
Acyclovir
5-bromouracil
Azidothymidine
Guanine nucleotide
Guanine nucleotide

6. What would happen if a C were substituted for base 16?

7. What would happen if a T were substituted for base 20?

8. What would happen if base 26 were deleted?

9. Does AUG occur only at the start of a protein? How can a ribosome "know" which AUGs represent start codons?

Fill-In Questions

Use the following choices to complete the statements below. Choices may be used once, more than once, or not at all.

amino acids
base analog
catabolic repression
degeneracy
DNA
DNA polymerase
induction
mRNA
negative (indirect) selection
positive (direct) selection
repression
ribosomes
RNA
RNA polymerase
semiconservative replication
tRNA

1. A cell that cannot make its own proline would be isolated by ______________________.
2. ______________________ inhibits gene expression and decreases the synthesis of enzymes.
3. The presence of glucose inhibits the *lac* operon by ______________________.
4. In the cell, protein synthesis occurs at ______________________.
5. Transfer RNA molecules carry ______________________ to the mRNA.
6. DNA polymerase copies ______________________ to make ______________________.
7. Most amino acids are coded by several alternative codons; this phenomenon is called ______________________.
8. Proofreading of new DNA is done by ______________________.
9. The enzyme responsible for transcription is ______________________.
10. A chemical that is similar to a nucleotide and mistakenly used by DNA polymerase is called a ______________________.

Short-Answer Questions

1. Differentiate between the terminator sequence and stop codons.

2. Differentiate between the start of mRNA transcription and the start codon for translation.

3. What is a constitutively expressed gene? A repressible gene? An inducible gene? A regulatory gene?

4. Is the *lac* operon an inducible or a repressible operon?

5. What happens when lactose binds the repressor for the *lac* operon?

6. What is the rate of spontaneous mutations? How do mutagens affect the mutation rate?

7. What would you find in a cell that lacked DNA ligase, that you would not find in a normal cell?

8. A strain of *E. coli* called χ1776 is often used in genetic engineering work. This strain cannot synthesize its own thymidine. Discuss what happens to these cells if a researcher doesn't give them thymidine.

9. The Ames test uses mutant bacteria that have lost their light-repair and histidine-synthesizing capabilities. Why are these mutants used?

10. Sometime in human history, a mutation changed the hemoglobin gene so that the resulting hemoglobin protein behaved differently than normal adult hemoglobin. This altered hemoglobin confers resistance to malaria and can lead to kidney failure. Read about sickle-cell disease on p. 222 in your textbook. Discuss whether this was a beneficial or a harmful mutation.

Hypothesis Testing

Students can post their arguments/evidence on an electronic bulletin board to arrive at a class conclusion.

Hypothesis: Antibiotic-resistance genes exist before bacteria are exposed to antibiotics.

Data/Observations: Design an experiment to test this hypothesis. Mutations occur randomly. After exposure to a lethal antibiotic there would not be enough time to produce resistance.

Conclusion: What results would lead you to accept the hypothesis? Briefly explain.

Study Questions

Microbiology: An Introduction, Sixth Edition				
Pages	Review Questions	Multiple Choice Questions	Critical Thinking Questions	Clinical Applications Questions
240–241	1–7, 9–10, 12	3–5	1–2	1–2

CD Activities

1. Do the Chapter 8 quiz.
2. Use the Genetics Unit to get hands-on practice with DNA.

Web Activities

1. Do the Chapter 8 quiz.
2. Read “Antibiotic Resistant Gonococci and Natural Selection.”
   a. How many years did it take for penicillin resistance to appear? From its first use in the 1940s to 1976.

   b. How many years did it take for tetracycline resistance to appear? From the time it was used to replace penicillin (1976) until 1986.

c. Will developing new antibiotics eliminate the problem of antibiotic resistance? No, it won't eliminate the problem. New antibiotics will provide alternative treatments until resistance to them is selected for in the bacterial population.

ANSWERS

Concept Map 4.1

1. Transcription produces RNA; this includes mRNA, rRNA, and tRNA.
2. The protein (peptide) is terminated at the stop codon. At the ribosome, each codon is translated into an amino acid.
3. The leading strand needs an RNA primer to give DNA polymerase a place to attach the first base; this strand is copied in one continuous chain. The lagging strand is copied in fragments as new pieces of the parent strand are exposed at the replication fork. Each fragment must start with an RNA primer.
4. The DNA unwinds, and the two complementary strands of DNA separate.

Definitions

1. leading strand
2. phenotype
3. RNA primer
4. RNA polymerase
5. mRNA
6. light-repair enzymes
7. frameshift mutations
8. cyclic AMP
9. base substitution
10. auxotroph

Practice with Base-Pairing and Mutations

1. 5′ ATGCCGTGTTAA
2. 5′ AUGCCGUGUUAA
3. met-pro-cys

4–7. AZT and acyclovir stop DNA synthesis. 5-bromouracil is inserted in place of thymine but pairs with guanine, which will cause a base substitution.

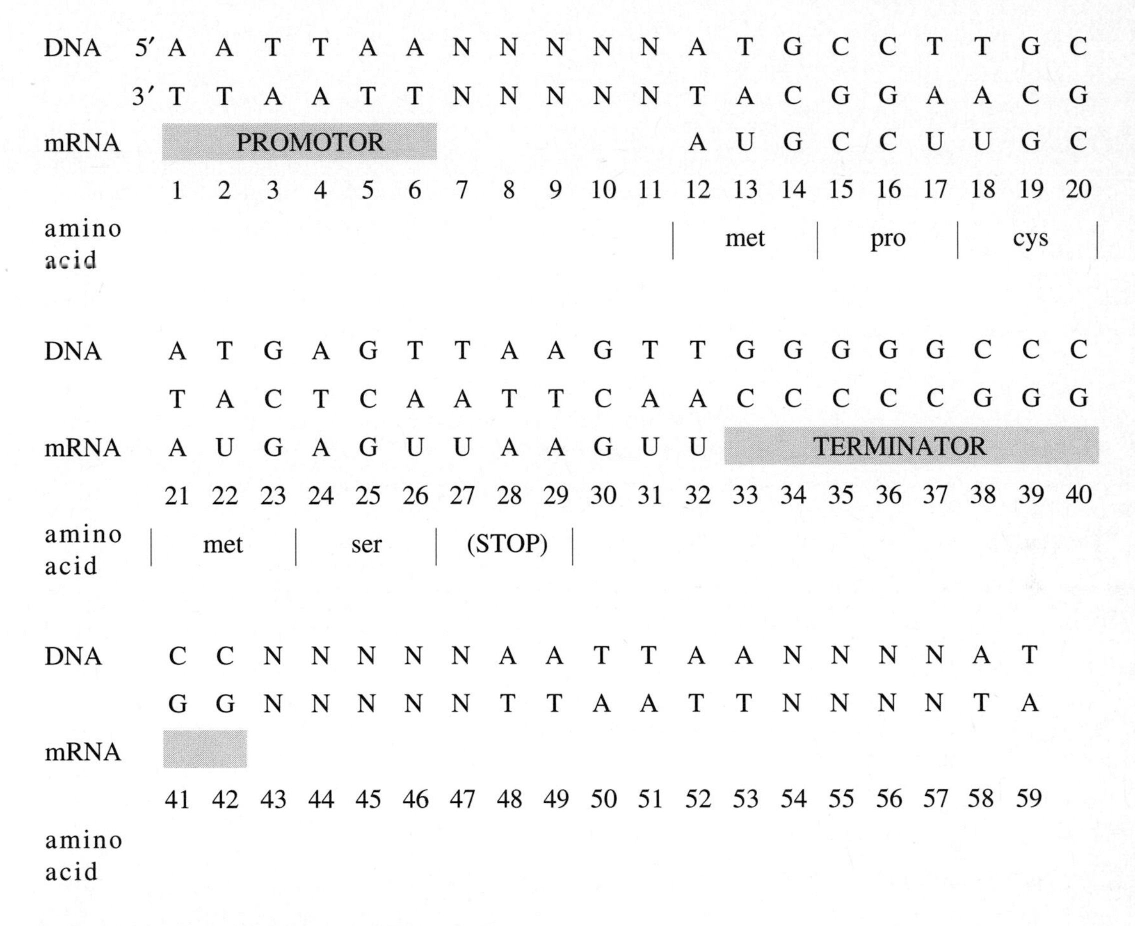

Practice with Base-Pairing and Translation

The answers to questions 1–4 are shown below:

5. No change; the codon CCC codes for proline.
6. Missense mutation; the amino acid would be changed to arginine.
7. Nonsense mutation; the peptide chain stops.
8. Frameshift mutation; the same amino acid sequence would be made, but it wouldn't stop at the correct place. Lysine would be added after serine. The cell would waste resources, adding amino acids until a stop codon was reached.
9. AUG can code for methionine in a protein. AUG is a start codon when it occurs downstream from the promoter.

Fill-In Questions

1. negative (indirect) selection
2. repression
3. catabolic repression
4. ribosomes
5. amino acids
6. DNA, DNA
7. degeneracy

8. DNA polymerase
9. RNA polymerase
10. base analog

Short-Answer Questions

1. RNA polymerase transcribes the stop codon, and transcription ends at the terminator. The stop codon signals the end of the peptide being synthesized at the ribosome because it does not code for an amino acid.
2. RNA polymerase binds to the promoter region of DNA; transcription starts at the start codon.
3. Constitutive genes are expressed all the time. Transcription of repressible genes is stopped when the gene products are not needed. Inducible genes are transcribed when their products are needed. Regulatory genes produce the repressor proteins.
4. The *lac* operon is induced by lactose.
5. Transcription of the structural genes occurs.
6. Once in 10^6 genes; mutagens increase the rate.
7. It would not attach the fragments of the lagging strand of DNA; you would find many small pieces of DNA in the cell.
8. These cells would not be able to grow because they could not make new DNA.
9. The Ames test is used to identify mutagens. Bacteria without light repair will be less likely to fix mutations, so the mutations resulting from the mutagens can be detected. The ability to synthesize histidine is the marker used to detect mutated cells.
10. It probably was a beneficial mutation because it allowed people to survive long enough to reproduce in malarial areas. Now, with mosquito-abatement techniques and antimalaria drugs, the sickle-cell gene has more harmful than beneficial effects.

Unit 5

GENETIC TRANSFER

Numerous bacterial variants are known which will grow in environments unfavorable to the parent strain . . . the variants arise spontaneously during growth under normal conditions.
—Howard Newcombe, 1949

Learning Objectives

After completing this unit, you should be able to:

★	1.	Compare the mechanisms of genetic recombination in bacteria.	p. 229
★	2.	Relate the mechanisms for genetic change to microbial evolution.	p. 229
★	3.	Describe the functions of plasmids and transposons.	p. 235
★	4.	Discuss how genetic mutation and recombination provide material for natural selection.	p. 237
	5.	Compare and contrast genetic engineering, recombinant DNA, and biotechnology.	p. 242
★	6.	Identify the roles of a clone and a vector in genetic engineering.	p. 243
	7.	Define restriction enzymes, and outline how they are used to make recombinant DNA.	p. 245
	8.	List the four properties of vectors.	p. 245
★	9.	Describe the use of plasmid and viral vectors.	p. 245
★	10.	Describe five ways of getting DNA into a cell.	p. 247
	11.	Describe how a gene library is made.	p. 248
	12.	Differentiate between cDNA and synthetic DNA.	p. 250
★	13.	Explain how each of the following is used to locate a clone: antibiotic resistance genes, DNA probes, gene product.	p. 250
	14.	List one advantage of engineering each of the following: *E. coli*, *Saccharomyces cerevisiae*, mammalian cells, plant cells.	p. 252
★	15.	List at least five applications of genetic engineering.	p. 253
★	16.	Outline genetic engineering with *Agrobacterium*.	p. 260
★	17.	List the advantages of, and problems associated with, the use of genetic engineering techniques.	p. 261
	18.	Describe the role of microorganisms in the production of industrial chemicals and pharmaceuticals.	p. 753

Reading

Chapter 8 (pp. 229–237), Chapter 9 (pp. 241–257, 260–263), Chapter 13 (pp. 372–373), and Chapter 28 (pp. 753–757).

SUGGESTED LABS FROM JOHNSON AND CASE, *LABORATORY EXPERIMENTS IN MICROBIOLOGY*, FIFTH EDITION

Exercise 28: Transformation of Bacteria
Exercise 30: Genetic Engineering

KEY TERMS AND CONCEPTS

Bacterial Gene Transfer
bacteriophage p. 234
conjugation p. 232
crossing-over p. 230
dissimilation plasmids p. 235
donor cell p. 230
F factor (fertility factor) p. 233
generalized transduction p. 234
genetic recombination p. 230
Hfr cell p. 233
insertion sequences (IS) p. 236
phage p. 234
plasmid p. 232
prophage p. 372
r-determinant p. 235
recipient cell p. 230
resistance transfer factor (RTF) p. 234
R factors (resistance factors) p. 235
specialized transduction pp. 235, 373
transduction p. 234
transformation p. 230
transposons p. 236

Biotechnology and Techniques
biotechnology p. 243
clone p. 243
colony hybridization p. 251
competence p. 232
complementary DNA (cDNA) p. 249
DNA fingerprinting p. 257
DNA probes p. 251
DNA sequencing p. 256
electroporation p. 247
exons p. 248
gel electrophoresis p. 256
gene library p. 248
gene therapy p. 256
genetic engineering p. 243
genetic screening p. 256
introns p. 248
microinjection p. 247
protease inhibitors p. 256
protoplast fusion p. 247
protoplasts p. 247
recombinant DNA p. 242
recombinant DNA technology p. 243
restriction enzymes p. 245
reverse transcriptase p. 249
RFLPs p. 256
shuttle vectors p. 245
Southern blotting p. 256
splicing p. 248
subunit vaccines p. 255
Ti plasmid p. 260
transformation p. 247

INTRODUCTION

Be sure to read the Study Outlines for Chapter 8, pp. 239–240, and Chapter 9, pp. 263–265.

Genetic recombination is the joining of DNA from different sources. In eukaryotic cells, genetic recombination is tied to reproduction and results in offspring with genes from two parent cells. In prokaryotes, reproduction is asexual and does not involve genetic recombination. However, genetic recombination does occur in prokaryotic cells: one cell may acquire DNA from another cell, which can result in the expression of new traits. The cell that has acquired new DNA is called a recombinant cell. There are three methods of genetic recombination in prokaryotic cells:

1. Conjugation requires cell-to-cell contact and the transfer of a plasmid from the donor cell, called an F^+ cell to a recipient, or F^-, cell. Recall from Unit 4 that plasmids are self-replicating circular DNA molecules. Plasmids often carry nonessential, but useful, genes such as those for antibiotic resistance.
2. Transformation occurs when a cell picks up DNA from the environment. The source of this DNA is a dead cell.
3. Transduction is the transfer of DNA from one cell to another by a virus.

Genetic recombination can also occur when transposons insert themselves into a chromosome. This insertion can inactivate a gene or bring new genes such as those for antibiotic resistance. Transposons may "jump out" of the chromosome at any time.

Biotechnology is the use of organisms to make a useful product. This includes the industrial production of penicillin and other microbial products. Selection of specific genetic recombinants has long been employed by humans in search of better products. Consider the breeding of dairy cows or racehorses. This kind of process is hit-or-miss because you can't be sure you'll get the desired product. In the 1980s, biologists developed methods of using the natural processes of transformation and transduction to insert desired genes into a recipient cell, thus ensuring the outcome. These modern molecular techniques are called genetic engineering.

Desired genes can be snipped from a chromosome using enzymes called restriction endonucleases or restriction enzymes. Restriction enzymes cut DNA somewhere in (endo = in) the nucleotide chain. There are several hundred restriction enzymes; each cuts at a specific base sequence. Enzymes were covered in Unit 3.

The DNA used to make a recombinant cell can come from three sources:

1. A DNA library
2. cDNA
3. Synthetic DNA (rarely)

Microorganisms, plant cells, and animal cells can be genetically engineered. There are a variety of reasons cells may be genetically engineered:

- To produce a product, such as an antibiotic, for manufacture.
- To make a disease- or pesticide-resistant plant.
- To cure a genetic disease.

See Figure 9.1 (p. 244), in your textbook for an overview of genetic engineering and examples of applications.

In the quote that opened this unit, Newcombe is talking about mutations, but genetic transfer also provides new variants. Genetic change provides bacteria with genetic diversity, which is necessary for natural selection and evolution, the subjects of Unit 6, Microbial Evolution.

VIDEO: GENETIC TRANSFER

Terms

The following new terms are introduced in this video:

horizontal gene transmission
phage conversion
transgenic
vertical gene transmission
virulence

Preview of Video Program

This video program starts with descriptions of the natural processes of conjugation, transformation, and transduction. In Unit 4, an animation showed that bacteria can transfer antibiotic resistance to other cells. This animation is used again in this video to demonstrate the process of bacterial conjugation. Recall from Unit 4 that Julian Davies and his colleagues were transferring DNA from soil bacteria to laboratory strains of bacteria. They were transferring the DNA using transformation. Transduction can lead to genetic change called phage conversion. *Corynebacterium diphtheriae* produces diphtheria toxin only when it is infected with a bacteriophage (virus). See the lysogenic cycle on p. 373 of your textbook.

These natural processes are used in genetic engineering to produce a specific product. The first example is from medicine. Like the Davies group, Maggie So uses transformation to study the products of specific genes. The second example is from agricultural microbiology. Victor Masona is trying to genetically engineer a virus-resistant plant. The goal of his work is to replace the current crop plants that are susceptible to cassava mosaic virus with the recombinant plants. Genetic engineering in plants is usually done with *Agrobacterium*, an unusual plant pathogen that inserts a tumor-inducing plasmid into its host plants. Masona is inserting a gene into *Agrobacterium* by electroporation; the *Agrobacterium* will then transfer the plasmid and new gene to its host plant cell, making a recombinant plant cell. The path to a useful product is quite long: The recombinant plant cells need to be cloned, then the cloned cells are separated and grown into plants. The first plants must be tested to ensure that they didn't acquire any undesirable traits. Then plants need to be grown to produce seeds for farmers.

Video Questions

1. Write definitions of the following terms as you watch the video: (As shown on slates in the video.)

   a. Transduction. Genetic material passes via a bacterial virus called a bacteriophage.

   b. Conjugation. Donor contacts recipient via conjugation tube known as a sex pilus and transfers DNA to recipient.

   c. Transformation. Bacteria bind DNA from their surroundings.

   d. Transposition. DNA that can hop to new location in a genome.

2. **Compare and contrast vertical and horizontal gene transfer.** Both transfer genes from one organism to another. In vertical transmission, genes are transmitted to a new generation. This occurs in humans when parent egg and sperm cells fuse to form a new offspring. In horizontal transmission, genes are transmitted to another member of the same generation. This occurs between bacteria by transformation, transduction, and conjugation.

3. What vector does Victor Masona use to engineer cassava plants? *Agrobacterium*

4. What is phage conversion? In lysogeny, the host cell acquires new characteristics from the viral genome.

5. Conjugative plasmids are transferred from one cell to another by conjugation. How can nonconjugative plasmids be transferred? Nonconjugative plasmids are mobilized by recombining with conjugative plasmids.

EXERCISES

Concept Map 5.1

This concept map shows terms related to industrial microbiology and biotechnology.

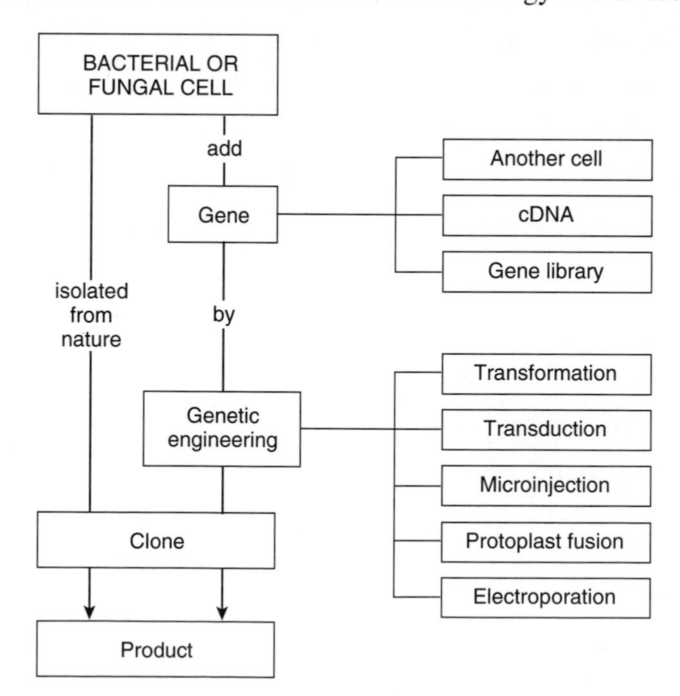

1. Where do the following terms fit in this map?
   a. Reverse transcriptase
   b. Horizontal transmission
   c. Virus or bacteriophage
   d. *Agrobacterium*
   e. Vector
2. If you genetically engineered a plant cell, what step(s) would you need to add?

Figure 5.1

Isolating a plasmid vector. Cut the three plasmid pieces on the solid lines and tape them together to form a circular molecule of DNA. In DNA synthesis, the 5′ end of one sugar is attached to the 3′ end of the preceding sugar. This plasmid carries the (shaded) genes for resistance to the antibiotics ampicillin and kanamycin.

Isolating a gene. Genes can be isolated or, if small enough, synthesized. Cut out the gene on the solid lines.

Restriction enzyme digestion. Your scissors are the restriction enzyme *Eco*R1, which cuts between the G and A in the sequence G↓AATTC on both strands of double-stranded DNA. The enzyme reads from the 5′ end of DNA. Note that the two strands of DNA are complementary, so you will have staggered (or sticky) ends after your cuts. Cut the plasmid and gene. Although enzymes work by trial and error to find their substrate, the arrows (↑) will help you locate the correct sequence.

Ligation. You can use tape in place of the DNA ligase that covalently joins pieces of DNA. Match the complementary bases of the staggered ends to tape the gene into the plasmid.

Voilà! A recombinant plasmid to insert into a cell.

Questions

1. How will you identify cells carrying the recombinant plasmid?
2. Cells carrying the original plasmid?
3. Cells without a plasmid?
4. Where was this technique used in the video?

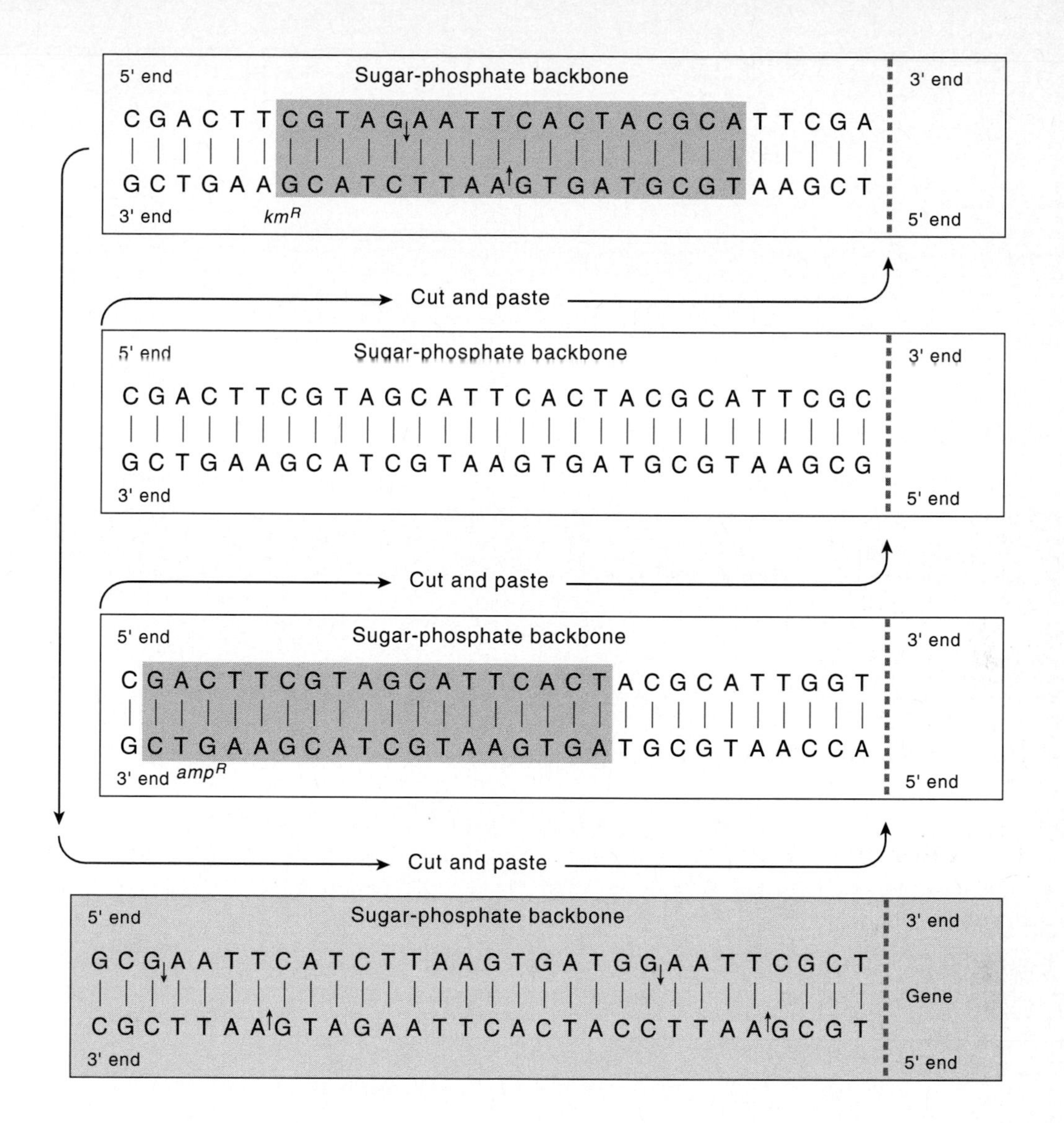
5' end
Sugar-phosphate backbone
3' end
C G A C T T C G T A G A A T T C A C T A C G C A T T C G A
G C T G A A G C A T C T T A A G T G A T G C G T A A G C T
3' end
kmR
5' end
Cut and paste
5' end
Sugar-phosphate backbone
3' end
C G A C T T C G T A G C A T T C A C T A C G C A T T C G C
G C T G A A G C A T C G T A A G T G A T G C G T A A G C G
3' end
5' end
Cut and paste
5' end
Sugar-phosphate backbone
3' end
C G A C T T C G T A G C A T T C A C T A C G C A T T G G T
G C T G A A G C A T C G T A A G T G A T G C G T A A C C A
3' end ampR
5' end
Cut and paste
5' end
Sugar-phosphate backbone
3' end
G C G A A T T C A T C T T A A G T G A T G G A A T T C G C T
Gene
C G C T T A A G T A G A A T T C A C T A C C T T A A G C G T
3' end
5' end

Figure 5.2

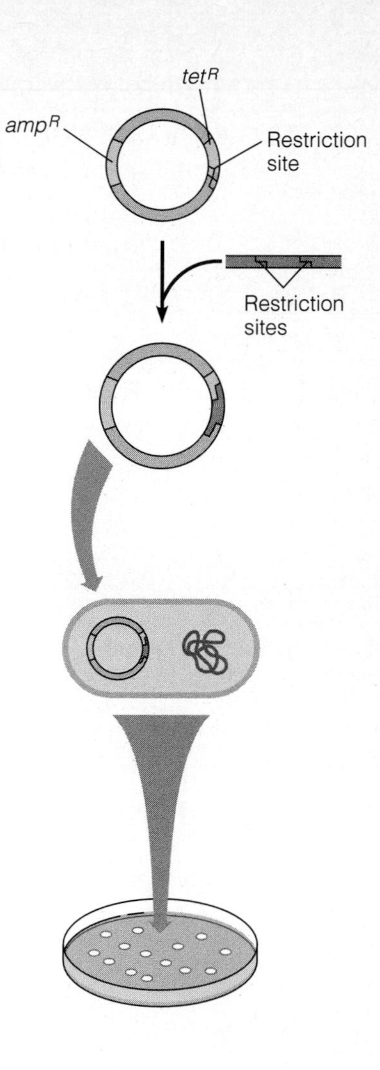

1. Color the plasmid and the new (foreign) gene different colors.
2. Label the recombinant plasmid alone and in the bacterial cell.
3. How will you locate cells that contain the recombinant plasmid?

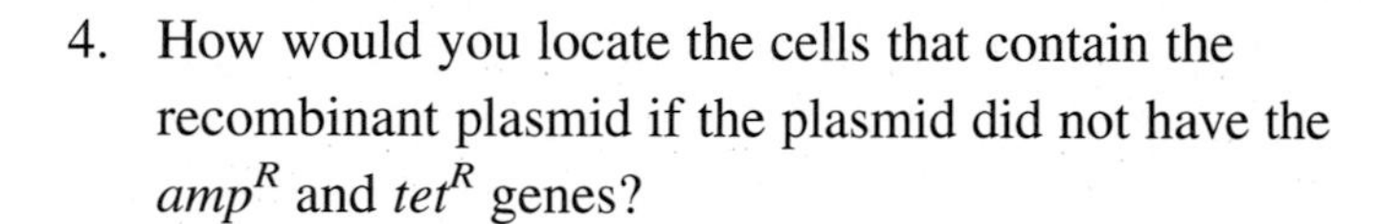

4. How would you locate the cells that contain the recombinant plasmid if the plasmid did not have the *amp*^{R} and *tet*^{R} genes?

Figure 5.3

Assume that you wanted to engineer a tomato plant to produce human milk protein, casein, and that the steps are shown in the figure.

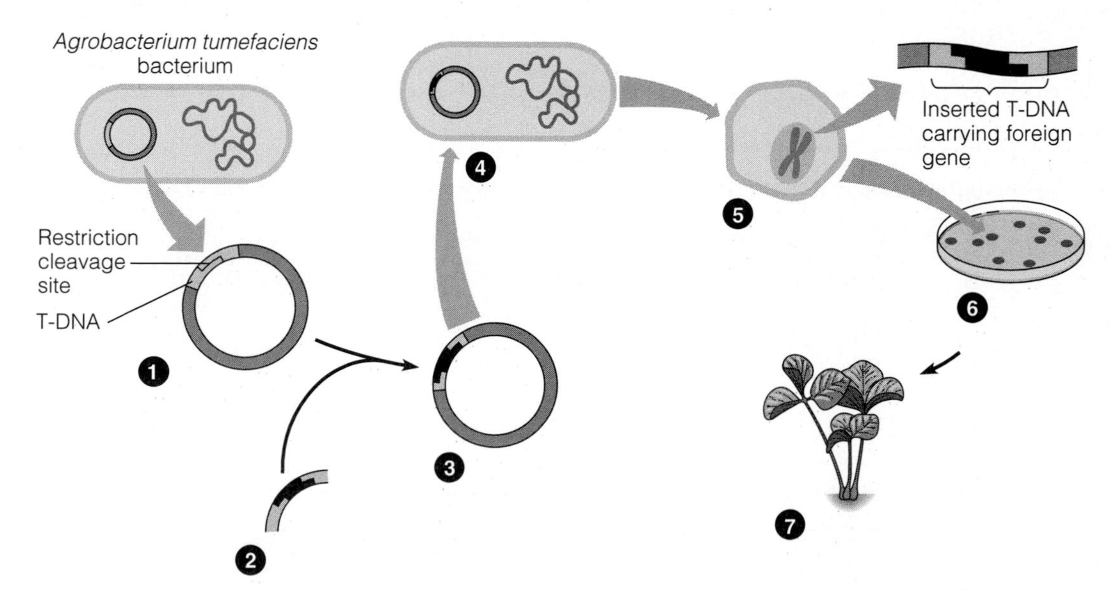

1. Label the following:
   a. Ti plasmid
   b. *Agrobacterium*
   c. Plant cell clones
   d. Milk protein gene
2. Show where each of the following occur in the figure:
   a. Restriction digest
   b. Transformation
   c. DNA ligase
3. How could you identify the desired clone?
4. What steps would change if you used microinjection? Electroporation?
5. The result of this experiment is (a) a human × tomato hybrid, (b) a human × bacteria hybrid, (c) a tomato × bacteria hybrid, or (d) a tomato plant?

Definitions

Match the following statements to words from Key Terms and Concepts:

1. Regions of eukaryotic chromosomes that code for proteins. ____________________
2. A technique used to separate DNA fragments. ____________________
3. A technique used to find a restriction fragment using a DNA probe. ____________________
4. A recombinant process that uses wall-less cells and polyethylene glycol.
   ____
5. A circular, self-replicating DNA molecule often used as a vector for genetic engineering.
   ____
6. Small pieces of DNA that can move from one region of a DNA molecule to another.
   ____
7. Extrachromosomal, circular, self-replicating DNA molecules that carries genes for resistance to antibiotics. ____________________
8. A bacterial cell in which the F factor has become integrated into the chromosome.
   ____

Matching Questions

Match the following choices to the statements below:

a. DNA ligase
b. DNA polymerase
c. Exons
d. Introns
e. Plasmid
f. Restriction enzymes
g. Reverse transcriptase
h. RNA polymerase

1. Enzymes that cut DNA to produce sticky ends. ____
2. The enzyme used to covalently join DNA fragments. ____
3. A shuttle vector. ____
4. An enzyme that copies DNA to make DNA. ____
5. The pieces of eukaryotic DNA that do not code for anything. ____
6. An enzyme that copies RNA to make DNA. ____

Fill-In Questions

Use the following choices to complete the statements below. Choices may be used once, more than once, or not at all.

conjugation	transduction	translation
transcription	transformation	

1. The genetic recombination process that requires cell-to-cell contact is ______________.
2. Certain viruses can move genes from one cell to another by ____________________.
3. The process by which a cell picks up naked DNA in solution is called ______________.
4. The process *Agrobacterium* uses to make a host plant produce growth hormones is ____________________.
5. The process *Agrobacterium* uses to transfer the Ti plasmid from one bacterium to another is ____________________.

Short-Answer Questions

1. Read about xanthan production on p. 254 in your textbook. How did the desired strain of *Xanthomonas* come about?
2. Many developing countries rely on rice and corn for food. There are many benefits to engineering these plants to produce more protein or to be drought resistant. Why can't *Agrobacterium* be used for this?
3. Why can't human DNA be placed directly into bacterial cells?
4. Describe how a cell can lose a plasmid but not a chromosome during cell division.
5. Why don't transposons make good vectors for genetic engineering?

A Bioethics Question

This is an important topic and can provide an opportunity for students to interact with each other while trying to resolve some of the ethical issues of genetic engineering. Students can post their arguments/evidence on an electronic bulletin board to arrive at a class conclusion.

Question: In the video, you heard that the Zimbabwe government is making laws regarding the release of genetically engineered organisms.

What are the benefits of genetic engineering? The risks? Students will have a wide variety of responses. Benefits should include providing disease-resistant crops, which would make more food available for people. Risks should include the possibility of a toxic product and alteration of the gene pool and, therefore, evolution.

Conclusion: What recommendations do you have regarding the release of genetically engineered organisms?

Study Questions

Microbiology: An Introduction, Sixth Edition				
Pages	**Review Questions**	**Multiple Choice Questions**	**Critical Thinking Questions**	**Clinical Applications Questions**
265–266	1, 3–6, 8–10	1–3, 5	1–2	2
240–241	8, 11, 13	1, 2, 6	3	
759	1			

CD Activities

1. Do the Chapter 9 quiz.
2. Use the Recombination and Cloning a Gene activities.

Web Activities

1. Do the Chapter 9 quiz.
2. Read "A flu from 1918 can make you sick today." Propose a mechanism by which a virus can have genes from two different viruses. The genes of two viruses replicating in one host cell combined.

ANSWERS

Concept Map 5.1

1. a. Used to make cDNA.
   b. Adding a gene to a cell.
   c. Transduction.

d. Transformation.
e. Transformation and transduction.

2. Start with a plant cell. Grow plants from cell clones.

Figure 5.1

After transformation, the cells are inoculated into petri plates containing nutrient agar (NA) without antibiotics and NA containing ampicillin (amp) and/or kanamycin (kan). Recombinant cells carrying the new gene will be resistant to ampicillin but not to kanamycin. You would identify the cells by their growth as shown below:

	Growth on			
Cells with	**NA**	**NA + amp**	**NA + kan**	**NA + amp + kan**
1. Recombinant plasmid	+	+	–	–
2. Original plasmid	+	+	+	+
3. No plasmid	+	–	–	–

4. In the video, Maggie So used this technique.

Figure 5.2

Check your answers to questions 1 and 2 with Figure 9.9 (p. 251) in your textbook.

3. Recombinant cells will be ampicillin resistant and tetracycline sensitive. These cells can be located by replica plating (Figure 8.18, p. 226, in your textbook).
4. Colony hybridization for presence of the new DNA (Figure 9.10, p. 252) or Western blotting (Figure 10.12, p. 285) for expression of a protein.

Figure 5.3

For questions 1 and 2, check your labels with Figure 9.16 (p. 261) in your textbook.

3. The clone could be detected by hybridization with a DNA probe in colony hybridization.
4. Microinjection and electroporation can be used to insert genes into the plant cell at step 5; transformation of *Agrobacterium* and by *Agrobacterium* would be eliminated.
5. d

Definitions

1. exons
2. gel electrophoresis
3. Southern blotting
4. protoplast fusion
5. plasmid
6. transposons
7. R factors
8. Hfr cell

Matching

1. f
2. a
3. e
4. b
5. d
6. g

Fill-In Questions

1. conjugation
2. transduction
3. transformation
4. transformation
5. conjugation

Short-Answer Questions

1. Mutation.
2. Normally, *Agrobacterium* only infects dicotyledons (broadleaf plants), not monocots.
3. Prokaryotic cells do not usually splice out introns.
4. Cell division occurs in conjunction with chromosome replication and segregation of the new chromosomes to each offspring cell. Plasmids replicate themselves, but their distribution in a cell is random and a plasmid may not be in the cytoplasm of one offspring cell.
5. Transposons are not stable; they can leave a chromosome at any time and move between species.

Unit **6**

MICROBIAL EVOLUTION

Microbes invented all of life's essential chemical systems, all of its rules for living and change.
—LYNN MARGULIS, 1982

LEARNING OBJECTIVES

	1.	Define taxon, taxonomy, and phylogeny.	p. 268
★	2.	Discuss the limitations of the two-kingdom system of classification.	p. 268
	3.	List the major characteristics used to differentiate the five kingdoms.	p. 269
★	4.	List the characteristics of the Bacteria, Archaea, and Eukarya domains.	p. 270
	5.	Differentiate among eukaryotic, bacterial, and viral species.	p. 273
	6.	Explain why scientific names are used.	p. 273
	7.	List the major taxa.	p. 274
	8.	Compare and contrast classification and identification.	p. 278
	9.	Explain the purpose of *Bergey's Manual.*	p. 278
	10.	Describe how staining and biochemical tests are used to identify bacteria.	p. 279
	11.	Differentiate between Western blotting and Southern blotting.	p. 282
	12.	Explain how serological tests and phage typing can be used to identify an unknown bacterium.	p. 282
	13.	Describe how a newly discovered microorganism can be classified by the following molecular methods: amino acid sequencing, DNA base composition, rRNA sequencing, DNA fingerprinting, PCR, and nucleic acid hybridization.	p. 284
	14.	Differentiate between a dichotomous key and a cladogram.	p. 290

READING

Chapter 10 and pp. 155–156.

SUGGESTED LAB FROM JOHNSON AND CASE, *LABORATORY EXPERIMENTS IN MICROBIOLOGY,* FIFTH EDITION

Exercise 32: Unknown Identification and *Bergey's Manual*

KEY TERMS AND CONCEPTS

amino acid sequencing p. 285
Animalia p. 270
antiserum p. 282
Archaea p. 271
bacterial species p. 276
binomial nomenclature p. 273

biochemical tests p. 279
cladograms p. 290
class p. 274
dichotomous keys p. 290
differential staining p. 279
division p. 274
DNA base composition p. 286
DNA fingerprinting p. 287
DNA probes p. 289
domain p. 274
enzyme-linked immunosorbent assay (ELISA) p. 283
eubacteria p. 271
eukaryotic species p. 274
family p. 274
five-kingdom system p. 269
Fungi p. 270
genus p. 273
kingdom p. 274
Monera p. 269
nucleic acid hybridization p. 289
order p. 274
phage typing p. 284
phylogeny p. 269
phylum p. 274
Plantae p. 270
polymerase chain reaction (PCR) p. 288
Prokaryotae p. 269
ribosomal RNA (rRNA) sequencing p. 287
serological testing p. 282
serology p. 282
slide agglutination test p. 282
Southern blotting p. 289
species p. 273
specific epithet p. 273
taxa p. 269
taxonomy p. 268
viral species p. 278
Western blotting p. 283

INTRODUCTION

Be sure to read the Study Outline for Chapter 10, pp. 291–292.

This unit focuses on the classification of organisms, which is based on their relatedness. All organisms are related by their evolution from a common ancestor. The evolutionary history of organisms is called phylogeny, and the tree of life shown in the video is a phylogenetic tree. Phylogeny is based on conservation of successful genes through successive generations. This means that organisms have genes that they inherited from their ancestors. The presence of *nearly identical* genes in many different organisms suggests that these organisms evolved from a common ancestor. For example, many organisms, from bacteria through humans, have a gene for cytochrome *c*. The genes and the cytochrome proteins for which they code are strikingly similar (see Figure 10.14, p. 287 of your textbook). This similarity is accounted for by what Charles Darwin called "descent with modification." Once life was established on Earth, different organisms arose from that original cell. In Unit 4 you learned that mutations constantly occur and that these mutations are the agents of change.

Classifying organisms can help us understand the origins of life. It also has immediate, practical applications. For example, you might want treatment for a microbial infection but can't find a drug that kills all kinds of microbes. Instead, you find drugs that are specific for bacteria, fungi, viruses, protozoa, or parasitic worms. Now you need to classify the microorganism. Even if you don't identify the organism to species, knowing whether it is a bacterium or virus will help you find an appropriate treatment. Read about a recent classification problem on p. 325 (Deuteromycota) of your textbook.

Organisms are classified in a hierarchy of taxa (categories) from the broadest, domain, to the most specific, species, based on degrees of genetic similarity between organisms. There are three domains: Bacteria, Archaea, and Eukarya. The accepted domain name is now Bacteria, although the domain name Eubacteria was in use when your textbook was published. The hierarchy is domain, kingdom, phylum (division), class, order, family, genus, species. Members of a species have the same major genetic traits. Similar species are grouped in a genus, similar genera in a family, and so on. In eukaryotic organisms, members of a species can reproduce with each other because they share these major genes.

Until the mid-1990s, a phylogenetic tree for bacteria could not be made because scientists had no way of determining which bacteria were more closely related. Now, however, ribosomal RNA (rRNA) sequencing is being used to determine relatedness between species of bacteria and archaea and to build a phylogenetic tree. The second edition of *Bergey's Manual of Systematic Bacteriology*, due to be published in 2000, will show the organization of bacteria and archaea into kingdoms, phyla, classes, orders, and families. The absence of a phylogenetic tree did not prevent researchers and medical microbiologists from identifying bacteria. Microbiologists working in clinical labs, in quality control in food manufacturing, or in water treatment labs need to identify microorganisms to trace an unwanted organism to its source or to verify the correct organism in a food manufacturing process. Bacteria, like other organisms, have been assigned to genera and given specific epithets for over 100 years, and identification is based on the presence of unique characteristics for each species.

Many eukaryotic organisms can be identified by looking at their phenotypic characteristics—that is, their macroscopic and microscopic features. However, bacteria look too similar to each other to rely on visual inspection; therefore, biochemical tests for enzymes are frequently used to identify bacteria. For example, recall from Unit 3 that all bacteria do not ferment glucose or lactose. Keep in mind that these enzymes are gene products like the external ears (pinnae) used to distinguish between a seal and a sea lion. These traits are inherited from the parent organism and developed during the phylogeny of the species. As we saw in Unit 4, genes may mutate, so the offspring may not always be identical to the parent. Some of the results of these changes are the subject of Unit 7, Microbial Diversity.

VIDEO: MICROBIAL EVOLUTION

Terms

The following new terms are introduced in this video:

hyperthermophiles
tree of life

Preview of Video Program

This video program starts in Yellowstone National Park, where, in the 1960s, Thomas Brock and his student Hudson Freeze discovered the thermophile *Thermus aquaticus*. Since then, microbiologists have found many bacteria that thrive at hot temperatures. The DNA polymerase of *T. aquaticus* is used in PCR because the enzyme is not inactivated by high temperatures (see Unit 7).

In the video, Karl Stetter, a microbiologist, collects a sample from a thermal vent at 90–95°C.

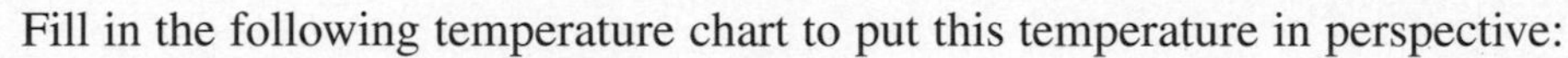

Fill in the following temperature chart to put this temperature in perspective:

- Home freezer temperature (~–10°C)
- Home refrigerator temperature (~7°C)
- Human body temperature
- Temperature at which water freezes
- Temperature of boiling water
- Highest temperature reached by Stetter's thermometer in the video (100°C)

Use Figure 6.1 (p. 155) of your textbook to find the following:

- Growth range of extreme thermophiles (hyperthermophiles)
- Growth range of mesophiles
- Growth range of psychrophiles
- Growth range of thermophiles

Stetter is collecting hyperthermophiles (also called thermoacidophiles), which grow at temperatures between 65°C and 113°C. The hyperthermophiles Stetter is collecting from Vulcano use hydrogen gas or hydrogen sulfide as their energy source. Recall from Unit 3 that these bacteria are chemolithotrophs, they live in an environment similar to the early Earth, and investigating them may provide information about the first life-forms. This will be discussed in Unit 7.

In the seventeenth century, Linnaeus classified the organisms that were known into two kingdoms, Plant and Animal. Subsequently, biologists filled in the kingdoms to show evolutionary relationships based on the fossil record. For example, mollusks must have evolved before mammals because no evidence of mammals occurs when fossil mollusks first appeared. Since then, new techniques—including the microscope—have revealed other organisms that don't fit the definitions of plants and animals. Bacteria don't look very different from each other, and a fossilized 3-billion-year-old cell can look like very common modern organisms. This doesn't mean, however, that they are the same organism or even closely related. Looks can be deceiving! Not too long ago, people thought whales were fish and koalas were bears, based on superficial observation.

In the 1970s, Carl Woese proposed a three-domain system of classification that is now widely accepted among biologists. This system groups organisms based on similar rRNA. Recall from Unit 4 that RNA is made of four nucleotide bases (A, C, G, and U) arranged in different combinations. This system appears "natural" because it is based on cells—the fundamental unit of life. Organisms in a domain are further divided into kingdoms based on degrees of relatedness. One way to look at relatedness is to look at genes that are shared by all organisms.

Woese chose an rRNA because all organisms have ribosomes. In the video, Woese is seen looking at autoradiographs of electrophoresis of rRNA from different organisms. Data collected from analyzing rRNA is confirming existing phylogenetic trees (which were based on the structures of living organisms as well as fossils) and allowing researchers to add bacteria and archaea to the tree of life.

Video Questions

1. How do the chemolithotrophs in the thermal vents get energy from H_2 and use sulfur instead of oxygen? These bacteria oxidize the hydrogen, transferring electrons to a coenzyme like NAD^+. The electrons are then used to generate ATP in the electron transport chain. Sulfur can be used as the final electron acceptor. Instead of aerobic respiration ($2H^+ + O_2 \rightarrow H_2O$), some of these bacteria use anaerobic respiration ($2H^+ + S^0 \rightarrow H_2S$).
2. Why can you assume that there are microorganisms in the sample Stetter collected from the vent? The water in the bottle is turbid.
3. Classification is based on ribosomal RNA. What is the function of ribosomes? Why do all organisms have ribosomes? Ribosomes are the sites of translation during protein synthesis. All organisms require structural and enzymatic proteins, which must be transcribed and translated from genes.
4. Why do we need to use rRNA sequencing; what's wrong with using observation as Linnaeus, Charles Darwin, and Louis Pasteur did? Many microorganisms are too similar in appearance to identify or classify them on this basis.
5. What percentage of a human's genes are the same as the genes of a whale, a dog, and a three-toed sloth? Why? All mammals share 97% of all genes; 3% is the small portion that differentiates species and organisms from each other. The genetic similarity can be accounted for by conservation of genes during the evolution of all organisms from a common ancestor.
6. What does Stetter mean when he says "the hotter the temperature, the deeper the branches go"? Stetter was talking about the branches on the phylogenetic tree. Figure 10.6 shows that thermoacidophiles (hyperthermophiles) are more primitive than the other Archaea, perhaps evolving from a thermophilic ancestor.
7. If all organisms evolved from a common ancestor, how do you account for the diversity of life on Earth? As Darwin observed, the descent from ancestors occurs with modification. Mutations do occur, and organisms with nonlethal mutations can survive. Some mutations may result in an organism looking a little different from its parent or using a different carbon or energy source than its parent.

EXERCISES

Concept Map 6.1

This map shows terms related to phylogeny.

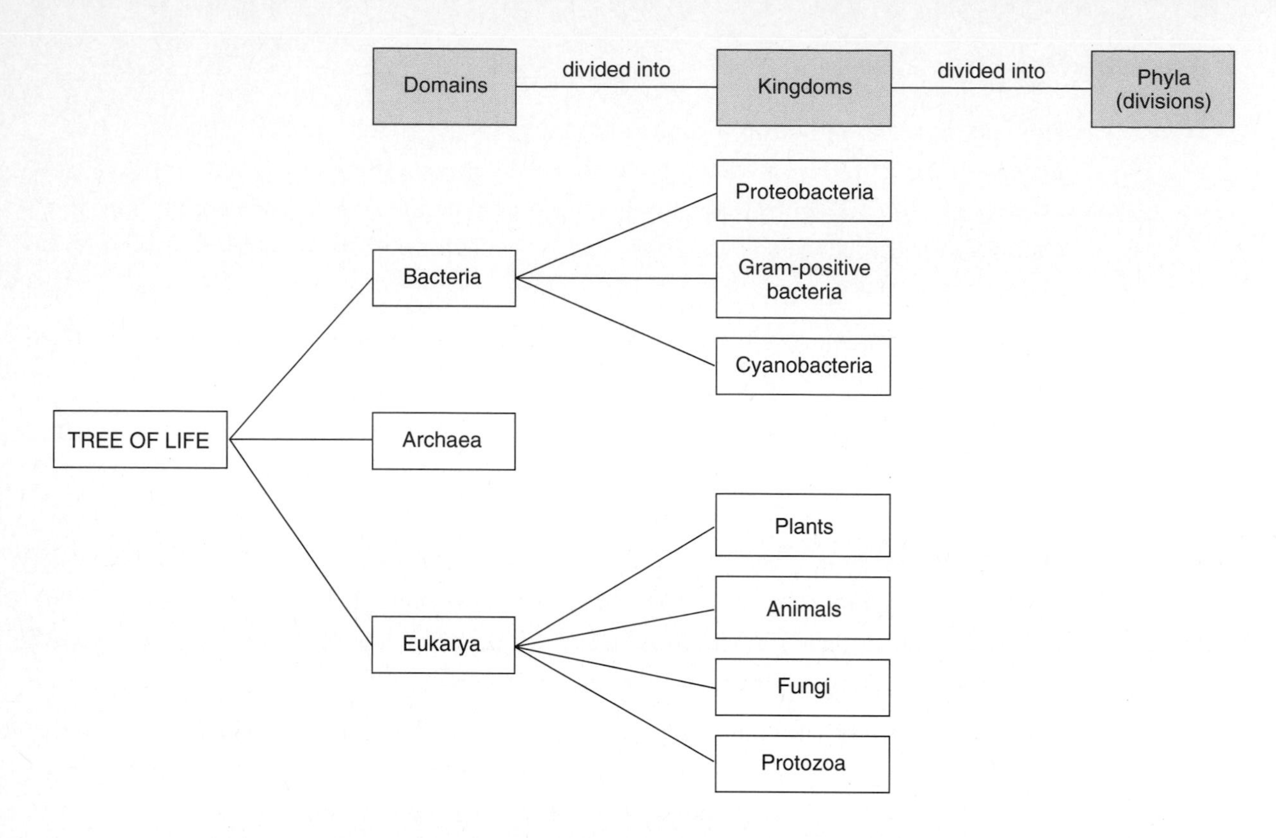

1. Add the kingdoms of Archaea to the map: Crenarchaeota (extreme thermophiles) and Euryarchaeota (methanogens and halophiles).
2. Where would you place algae in this concept map?
3. Why isn't this map a cladogram?

4. Using the following information, add a proposed classification for *Escherichia coli* to this map.

   E. coli bacteria have prokaryotic cells with ester-linked phospholipids in the plasma membrane; they have gram-negative cell walls. Based on rRNA sequencing, *E. coli* is in the gamma division of the Proteobacteria, along with several phototrophs and *Legionella*, Vibrionaceae, and Pasteurellaceae. *E. coli* is a chemoorganotroph along with *Legionella*, *Vibrio*, and *Pasteurella*. *E. coli* is a facultative anaerobe in the Enterobacteriales. This order includes the Vibrionaceae, Pasteurellaceae, and Enterobacteriaceae. *E. coli* is oxidase-negative and has peritrichous flagella. The Enterobacteriaceae include the genera *Salmonella*, *Shigella*, *Yersinia*, and *Escherichia*. *Bergey's Manual* lists the following species of *Escherichia: E. blattae, E. coli*, *E. fergusonii*, *E. hermannii*, and *E. vulneris*.

Additional information on selected genera:

	Escherichia	*Legionella*	*Pasteurella*	*Vibrio*
Cell wall type	Gram-negative	Gram-negative	Gram-negative	Gram-negative
Oxidase	Negative	Negative	Positive	Positive
Flagella	Peritrichous	Polar	Nonmotile	Polar
O_2 requirement	Facultative anaerobe	Aerobe	Facultative anaerobe	Facultative anaerobe

Figure 6.1

1. Place the following terms in their correct places in the figure:

a. Animals	f. Fungi	k. Methanogens
b. Archaea	g. Gram-negative bacteria	l. Mycoplasmas
c. Cyanobacteria	h. Gram-positive bacteria	m. Plants
d. Bacteria	i. Green nonsulfur bacteria	n. Protists
e. Eukarya	j. Halophiles	o. Thermoacidophiles

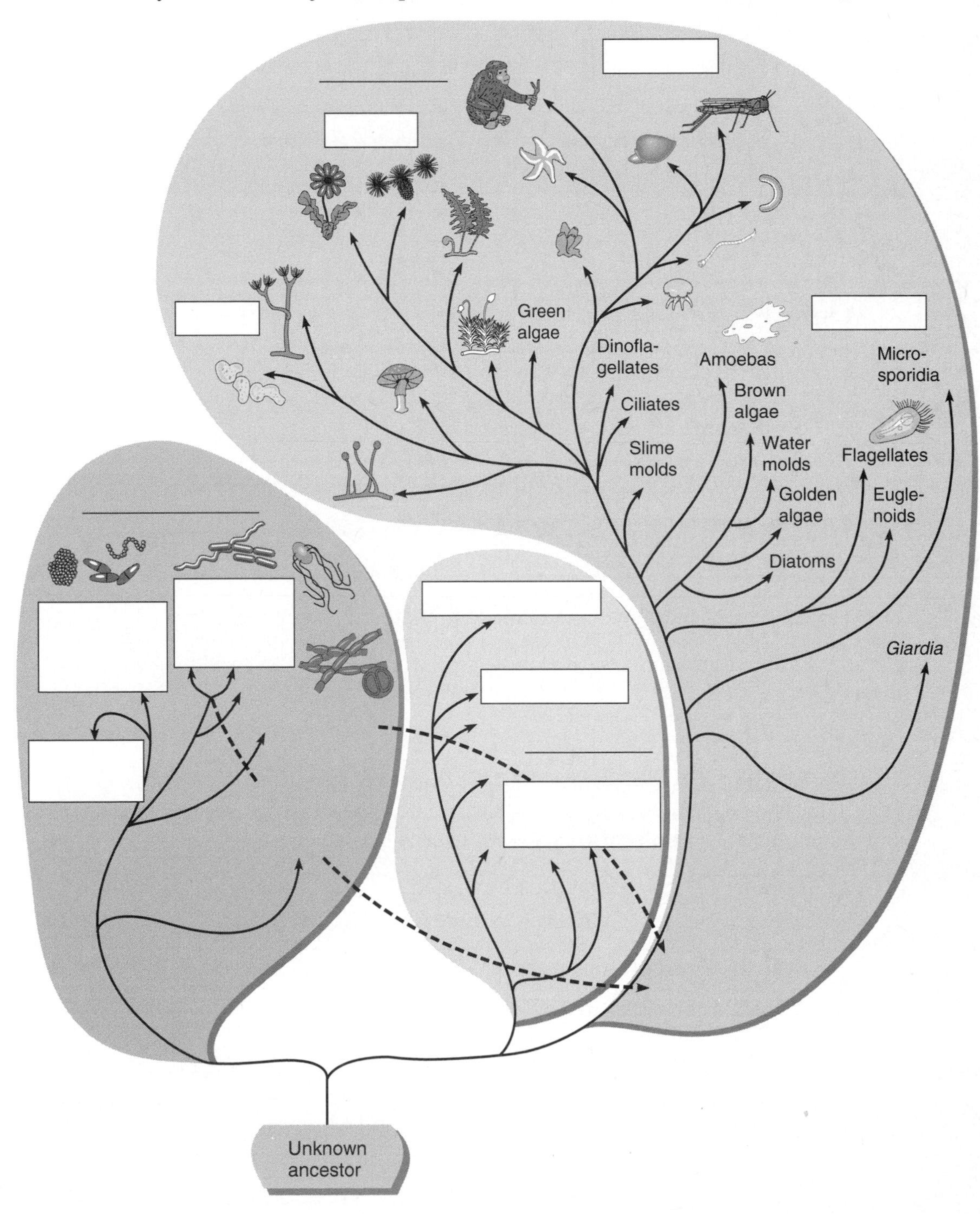

2. What do mycoplasma look like? Do they cause any diseases? (*Hint:* Look them up in the index to your textbook.)

3. Show the possible origins of the following eukaryotic organelles: mitochondria and chloroplasts.

4. The following organisms are used as examples in the video series. Add them to Figure 6.1 in their proper places:

   a. Humans
   b. *Escherichia coli*
   c. *Staphylococcus aureus*
   d. Methanotrophs (Unit 8)
   e. *Thermotoga maritima*
   f. *Trichomonas* Unit 7

5. Use your textbook to find where *E. coli* is found in nature. Where is *S. aureus* found?

Definitions

Match the following statements to words from Key Terms and Concepts:

1. The evolutionary history of a taxon. ____________________
2. The Gram stain is an example. ____________________
3. Testing based on metabolic characteristics. ____________________
4. Identification based on the presence of antigens. ____________________
5. The tree of life showing evolutionary relationships is an example. ____________________
6. Evidence that mitochondria arose from bacterial cells is provided by this test.

   ____

Matching Questions

Match the following choices to the statements below:

a. Animalia
b. Fungi
c. Gram-negative bacteria
d. Gram-positive bacteria
e. Plantae
f. Protista

1. A multicellular organism whose cells have nuclei and no cell walls. ____
2. A unicellular organism that is photosynthetic, has a nucleus, and lacks a cell wall. ____
3. A unicellular organism that lacks a nucleus and has a thick peptidoglycan cell wall. ____

4. A multicellular organism that is photosynthetic with nucleated cells and cellulose cell walls. ____

5. A photosynthetic, unicellular organism that lacks a nucleus and has a thin peptidoglycan cell wall. ____

Relatedness Questions

Read about the following organisms in your textbook to answer the questions below:

Escherichia (pp. 296–297, 302–303)
Halobacterium (pp. 296–297, 314)
Penicillium (p. 329)
Pseudomonas (pp. 296–297, 299–300)
Staphylococcus (pp. 296–297, 308)

1. What characteristics do *Escherichia* and *Staphylococcus* have in common? How do they differ?

2. What characteristics do *Escherichia* and *Pseudomonas* have in common? How do they differ?

3. What characteristics do *Escherichia* and *Penicillium* have in common? How do they differ?

4. What characteristics do *Escherichia* and *Halobacterium* have in common? How do they differ?

5. What characteristics do *Escherichia* and humans have in common? How do they differ?

Fill-In Questions

Use the following choices to complete the statements below. Choices may be used once, more than once, or not at all.

Bergey's Manual	family	*Origin of Species*	species
class	genus	phylum	strain
DNA	order	RNA	

1. Class, order, ______________________, genus, species.
2. In the name *Yersinia pestis*, *Yersinia* is the ______________________.
3. The name *E. coli* O157:H7 designates a bacterial ______________________.
4. The reference for bacterial taxonomy is called ______________________.
5. The G+C ratio is determined by chemical analysis of ______________________.

Short-Answer Questions

1. Why is the two-kingdom system used in the nineteenth century inadequate today?

2. Why was the two-kingdom system accepted for so long?

3. Why don't biologists just add a sixth kingdom for the Archaea; what is the value of reorganizing the taxonomic hierarchy to include the new taxon, domain?

4. One hundred fifty years ago, biologists classified fungi as plants. Now biologists believe that fungi are more closely related to animals than to plants. What characteristics do fungi and plants share? Fungi and animals? What evidence would show that fungi are more closely related to animals?

5. From the discovery of bacteria through the 1850s, biologists classified them as animals. What characteristics do bacteria and animals share? From the 1850s to 1969, biologists classified bacteria as plants. What characteristics do bacteria and plants share? What discovery in the 1960s led to bacteria being placed in their own kingdom? Why have biologists had so much trouble classifying bacteria?

6. The Kingdom Protista is sometimes described as "artificial" and "a junk drawer." Provide a rationale for these statements.

7. Purple phototrophic bacteria and many chemotrophic bacteria such as *Escherichia* have similar rRNA sequences, leading to the hypothesis that the phototrophs gave rise to the chemotrophs in this group. Provide a mechanism for this hypothesis.

8. Why did Woese believe that organisms with similar rRNA are more closely related?

Hypothesis Testing

This question provides an opportunity for students to interact with each other. They can post their arguments/evidence on an electronic bulletin board to arrive at a class conclusion.

Hypothesis: All organisms descended from earlier organisms.

Data/Observations: What evidence supports this hypothesis?

Conclusion: From these data, do you accept or reject the hypothesis? Briefly explain.

Study Questions

***Microbiology: An Introduction,* Sixth Edition**				
Pages	**Review Questions**	**Multiple Choice Questions**	**Critical Thinking Questions**	**Clinical Applications Questions**
292–294	All	All	All	All

CD Activities

1. Do the Chapter 10 quiz.
2. Use the *Bacteria ID* CD to practice using biochemical tests to identify bacteria. Do the problems on pp. 8–13 in the Guide at the end of your textbook.

Web Activities

1. Do the Chapter 10 quiz.
2. Read "Why Microbiologists Study Termites."
   a. How does *Mixotricha* support the hypothesis that eukaryotic flagella evolved from prokaryotic cells? The close relationship between the protozoan and the bacteria is startling.

   b. What additional evidence would you need to accept this hypothesis? Eukaryotic flagella with DNA and/or prokaryotes with 9+2 flagella.

Using a Key

A dichotomous key to the bacteria mentioned in *Unseen Life on Earth* is in Appendix A. Complete the key for identifying these bacteria.

ANSWERS

Concept Map 6.1

1.

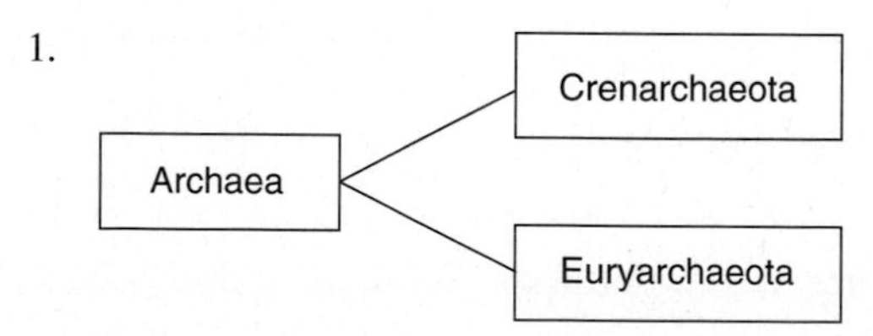

2. Algae should be a group of eukarya. Note that algae and protozoa are not kingdoms.

3. In a cladogram, the length and position of the connecting lines are used to show relatedness. All of the lines are not the same length and do not branch from the same point.

4.

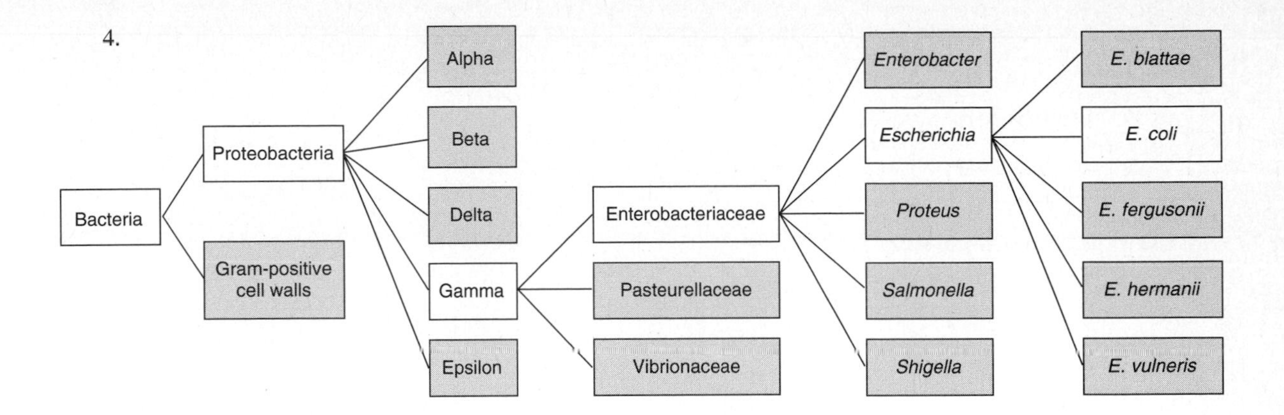

Figure 6.1

1. Check your answers against Figure 10.2 (p. 272) in your textbook.
2. You can read about mycoplasma on p. 306.
3. Follow the red lines in Figure 10.2.
4.

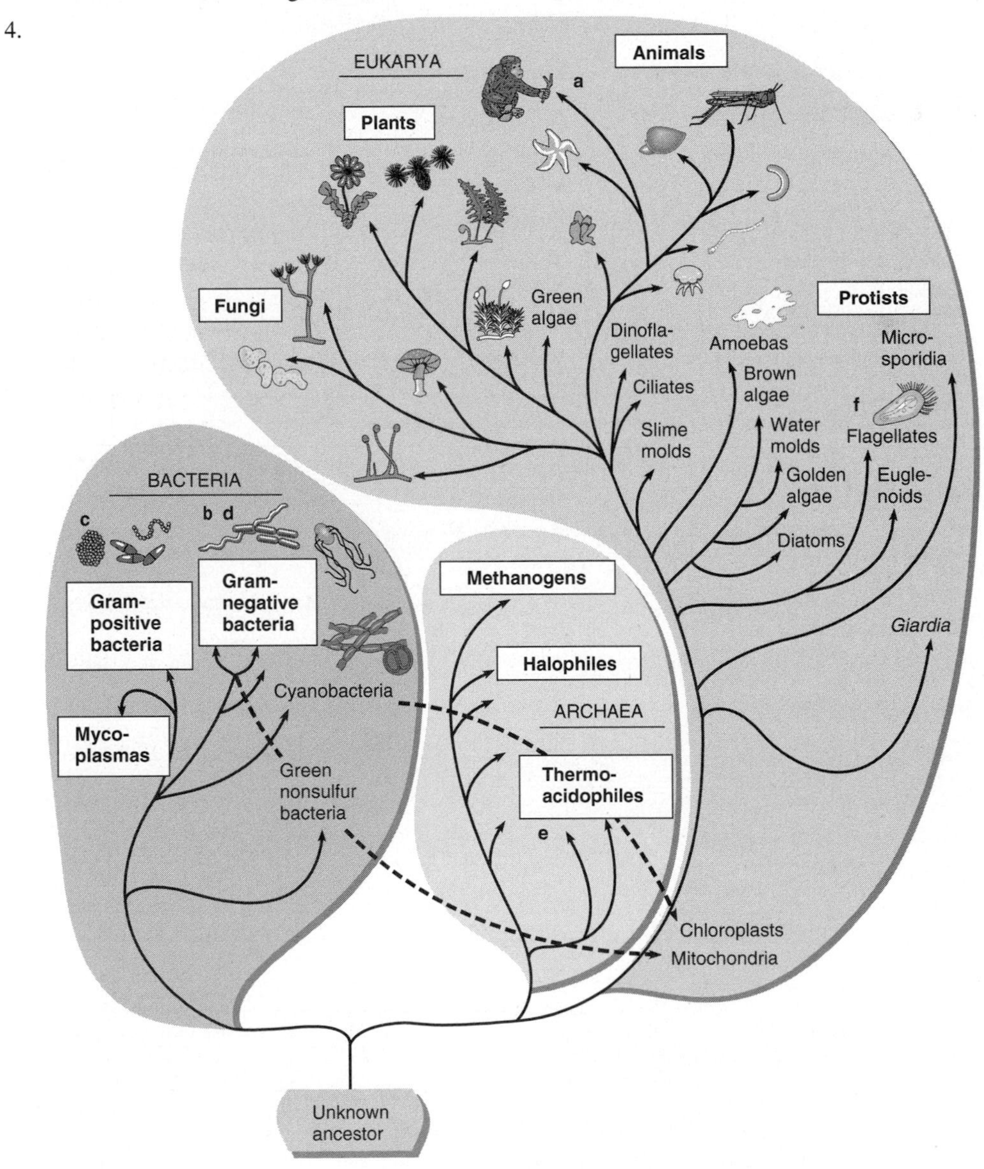

5. *E. coli* is found in vertebrates' intestines. *S. aureus* is found on human skin.

Definitions

1. phylogeny
2. differential staining
3. biochemical tests
4. serological testing
5. cladograms
6. rRNA sequencing

Matching Questions

1. a
2. f
3. d
4. e
5. c

Relatedness Questions

As living organisms, all of these organisms have many of the same metabolic enzymes, all have DNA, and all carry out protein synthesis.

1. They are both bacteria; therefore, they have ester-linked lipids in their membranes, peptidoglycan cell walls, and the other characteristics of bacteria. They are in different kingdoms because they differ in the structure and chemical composition of their walls: *Escherichia* is gram-negative and *Staphylococcus* is gram-positive.
2. They are both bacteria; therefore, they have ester-linked lipids in their membranes, peptidoglycan cell walls, and the other characteristics of bacteria. They are also in the same kingdom, because they share the gram-negative cell wall structure. These two organisms are in different genera because they have different oxygen requirements and flagellar arrangements, among other characteristics.
3. They are both living organisms composed of cells. However, they are in different domains because their cell structures differ.
4. They are both living organisms composed of cells. However, they are in different domains because their cell structures differ.
5. They are both living organisms composed of cells. However, they are in different domains because their cell structures differ.

Fill-In Questions

1. family
2. genus
3. strain
4. *Bergey's Manual*
5. DNA

Short-Answer Questions

1. Until the invention of the microscope in 1650, people could only classify what they could see; thus, the plant and animal kingdoms were acceptable.
2. It wasn't until 1937 that biologists realized that bacteria have prokaryotic cells—that is, they lack a nucleus. At this point, classifying them with eukaryotic plants was no longer acceptable.

3. Biologists generally look for natural groupings of organisms. Adding a sixth kingdom doesn't simplify learning the characteristics; instead, it would force us to memorize a list without looking for similarities.

4.

Characteristic	Fungi	Plant	Animal
Cell type	Eukaryotic	Eukaryotic	Eukaryotic
Nucleus	Yes	Yes	Yes
Cell wall	*Yes*	*Yes*	No
Motility	*No*, for most	*No*	Yes
Glycolysis	Yes	Yes	Yes
Krebs cycle	Yes	Yes	Yes
ETC	Yes	Yes	Yes
DNA	Yes	Yes	Yes
DNA & Protein synthesis	Yes	Yes	Yes
Photosynthesis	No	Yes	No
Nutritional type	Absorptive chemoorganotroph	Photoautotroph	Ingestive chemoorganotroph

The italicized items in the table influenced early taxonomists to include fungi in the plant kingdom; rRNA sequencing is providing the evidence that fungi and animals are related.

5.

Characteristic	Bacteria	Animal	Plant
Cell type	Prokaryotic	Eukaryotic	Eukaryotic
Cell wall	*Yes*, peptidoglycan	No	*Yes*, cellulose
Motility	*Yes*	*Yes*	No
Glycolysis	Yes	Yes	Yes
Krebs cycle	Yes	Yes	Yes
ETC	Yes	Yes	Yes
Anaerobic growth	Some	No	No
Ribosomes	70S	80S	80S
Photosynthesis	Some	No	Yes
Nutritional type	Varied; not ingestive	Ingestive chemoheterotroph	Photoautotroph

The italicized items in the table influenced early taxonomists to include bacteria in the plant kingdom. The defining of prokaryotic cells was significant in determining that bacteria were not plants. In 1961, Roger Stanier and Cornelis van Niel defined a prokaryotic cell by its lack of a nucleus. Bacteria are at the limit of resolution for most light microscopy, so early biologists were relying on the limited structural details they could see.

6. The kingdom contains a wide variation in organisms with no particular unifying theme. It includes very primitive chemoorganotrophs like *Giardia*, multicellular photoautotrophs like the brown algae, and funguslike slime molds.
7. One possible mechanism is that when an ancestral cell reproduced, some of its offspring cells lost one or more genes for photosynthesis but could survive if organic molecules were in the environment. Remember that photosynthetic organisms also have enzymes for metabolizing the organic molecules they make (i.e., glycolysis). The loss of genes and, therefore, abilities is called degeneracy.
8. The genes for rRNA are inherited. Organisms that diverged long ago from a common ancestor could have acquired different mutations in the gene. The closer organisms are, the less likely they are to have different mutations (and genes).

Unit **7**

MICROBIAL DIVERSITY

It must, however, be remembered that the functions of the nonpathogenic organisms in the economy of nature are as yet but very imperfectly understood, and as far as these functions have been investigated they do not yield in point of importance to those of the most virulent pathogenic forms.
—GRACE FRANKLAND, 1888

LEARNING OBJECTIVES

★ 1. Outline the PCR, and provide an example of its use. p. 258
★ 2. Compare the use of PCR and culture methods for analyzing natural populations of microorganisms. p. 316
3. Classify microbes into five groups on the basis of preferred temperature range. p. 154
4. Identify how and why the pH of culture media is controlled. p. 154
5. Explain the importance of osmotic pressure to microbial growth. p. 154
6. Provide a use for each of the four elements (carbon, nitrogen, sulfur, and phosphorus) needed in large amounts for microbial growth. p. 158
7. Explain how microbes are classified on the basis of oxygen requirements. p. 158
8. Identify ways in which aerobes avoid damage by toxic forms of oxygen. p. 158
9. Define extremophile, and identify two "extreme" habitats. p. 714
10. Outline the carbon cycle, and explain the roles of microorganisms in this cycle. (Also in Unit 8, which covers the nitrogen and sulfur cycles.) p. 716
11. Identify where microorganisms can be found in the ground. p. 716
12. Describe how an ecological community can exist without light energy. p. 723

READING

Chapter 5 (pp. 140–142), Chapter 6 (pp. 154–162), and pp. 81, 258–259 (PCR), 288 (PCR), 316–317, 714, 716–717, 723 (Life Without Sunshine), and 725–726. Skim Chapter 11.

SUGGESTED LABS FROM JOHNSON AND CASE, *LABORATORY EXPERIMENTS IN MICROBIOLOGY*, FIFTH EDITION

Exercise 29: DNA Fingerprinting
Exercise 35: Phototrophs: Algae and Cyanobacteria

Key Terms and Concepts

acidophiles p. 157
aerotolerant anaerobes p. 161
autotrophs p. 140
benthic zone p. 726
bioluminescence p. 726
carbon cycle p. 716
catalase p. 160
chemoautotrophs p. 142
chemoheterotrophs p. 142
chemotrophs p. 140
DNA fingerprinting p. 258
endoliths p. 723
extreme halophiles p. 158
extreme thermophiles p. 157
extremophiles p. 714
facultative anaerobes p. 160
facultative halophiles p. 158
heterotrophs p. 140
hyperthermophiles p. 157
limnetic zone p. 726
littoral zone p. 726
maximum growth temperature p. 155
mesophiles p. 155
microaerophiles p. 161
minimum growth temperature p. 155
nitrogen fixation p. 158
obligate aerobes p. 160
obligate anaerobes p. 160
obligate halophiles p. 158
optimum growth temperature p. 155
organic growth factors p. 162
peroxidase p. 161
peroxide anion p. 160
photoautotrophs p. 140
photoheterotrophs p. 141
phototrophs p. 140
phytoplankton p. 726
plasmolysis p. 157
polymerase chain reaction (PCR) p. 258
primary producers p. 723
profundal zone p. 726
psychrophiles p. 155
superoxide dismutase (SOD) p. 160
symbiosis p. 158
thermophiles p. 155
trace elements p. 160

Introduction

Be sure to read the Study Outlines for Chapter 6, pp. 177–178, and Chapter 11, pp. 317–318.

The diversity of microbial life is greater than all other species of life put together. The existence of these different organisms is accounted for by mutations. Recall from Unit 3 that most bacteria grow by binary fission. Before binary fission occurs, the cell's chromosome is copied so that each new cell can get a copy of all of the parent cell's genes. The replication process is fairly accurate; however, as you learned in Unit 4, some mistakes or mutations do occur. Nonlethal mutations may go unnoticed until the environment changes. A change that allows improved survival of the mutant cell will allow that organism to grow and populate the area. Nonmutated cells may still get enough nutrients to grow slowly—or they may not be able to survive.

Since Pasteur's first cultures, microbiologists have been studying microorganisms by growing or culturing them in the laboratory. Different culture media are used to provide nutrients the organisms would have in the wild. Culturing organisms provides information on their growth rates at different temperatures. Cultures are also used to study metabolic pathways and how microorganisms are using different chemicals, such as cellulose, sulfur, and even toxic wastes. Most cultures are grown as pure cultures; consequently, they don't provide any information on how one microbial species interacts with another species.

For over 100 years, microbiologists and ecologists have realized that the activities of microbes were important in nature. However, until recently, the only microbes that were studied were those that could be cultured in a laboratory. Most microbes haven't been cultured either because (1) they are part of an ecological community in which each organism is dependent on another or (2) we don't know their nutritional requirements.

The polymerase chain reaction (PCR) technique can be used to determine the presence of microorganisms that haven't been cultured. DNA can be extracted from soil, and a primer for a gene that (a) all organisms have or (b) only a select group possesses is added. The primer will hybridize with complementary sequences on DNA from the soil sample, and that soil DNA will be amplified in the PCR. The amplified DNA can be detected by gel electrophoresis. In (a), DNA from many different organisms will be obtained. In (b), the presence of amplified DNA means that particular organisms are present.

The work of Schmidt, Pace, Nelson, and Woese is confirming that the functions of the nonpathogenic organisms are important. Some of these important functions of microbes in nature will be the subject of Unit 8, Microbial Ecology.

VIDEO: MICROBIAL DIVERSITY

Terms

The following new terms are introduced in this video:

agarose gel
Proteobacteria
shotgun DNA sequencing

Preview of Video Program

This video program provides an overview of the diversity of microorganisms.

Tom Schmidt shows video micrographs of contents from the gut of a termite in which protozoa and prokaryotes are visible. However, as you learned in Unit 6, the prokaryotes look too much alike to identify them. Schmidt demonstrates the use of a DNA fingerprinting technique to learn about these bacteria.

DNA Fingerprinting

a. Cells from a sample such as the microbial community in a termite gut are collected.
b. DNA is extracted from the cells. The cells are lysed to release their DNA. Alcohol is added to the solution to precipitate the DNA. The DNA can be collected by spooling it on a glass rod; this is shown in a later image (at ~20 min. in the video).
c. Regions of the DNA are amplified. A primer for a region of rRNA that is found in all organisms is used for the PCR reaction, then the DNA is amplified for about 30 cycles.
d. Electrophoresis is used to separate the pieces of DNA. A blue tracking dye is added to the mixture, which is then loaded into wells in a slab of agarose gel. Agarose is highly purified agar. When the agarose is exposed to an electric current, the pieces of DNA migrate through the agarose. To which pole (positive or negative) do the pieces of DNA move? + ________ Which pieces move faster, smaller or larger ones? smaller ________
e. The DNA is stained to visualize it. Schmidt adds ethidium bromide, which combines with the DNA, and places the gel on an ultraviolet (UV) light. The ethidium fluoresces with UV light, and pink bands of DNA are visible.

The next step is to sequence the DNA to look for different organisms. In the video, this is shown in Karen Nelson's lab at The Institute for Genomic Research (TIGR). The amplified pieces of DNA are cut into smaller pieces with restriction enzymes, and an automated DNA sequencer is used to sequence the As, Ts, Cs, and Gs. Then the researcher must evaluate the products of the sequencer to put the pieces in the correct order. This is called shotgun sequencing. There may be missing segments that will still have to be sequenced one piece at a time.

The video returns to Schmidt's lab to look at a phylogenetic tree being made with information from sequenced DNA.

In Unit 6, you saw that Carl Woese used rRNA base sequences to determine evolutionary relationships between organisms. In this video program, you see Norm Pace use rRNA to detect unknown, uncultured bacteria in natural samples. Pace uses a primer for the rRNA genes for PCR amplification of DNA in samples from hot springs.

DNA Cloning

a. DNA is isolated and extracted as in (a) and (b) above.
b. Regions of the DNA are amplified by PCR using a primer for an rRNA gene. The DNA is amplified for about 30 cycles. If there was one rRNA gene in the sample, how many copies of that rRNA gene will be made? 1.07×10^9
c. PCR products and plasmid vectors are digested with a restriction enzyme. The PCR products are then ligated into the vectors.
d. The recombinant vectors are used to transform *E. coli* to create a gene library (Figure 9.6, p. 249). As the bacteria reproduce, the DNA fragments are also replicated, or cloned.
e. DNA is extracted from the *E. coli* cells for sequencing.

Video Questions

1. Why does a microbiologist study termites? Termites house a variety of microbes. These microbes digest cellulose and provide other nutrients for the termites.

2. How is PCR used to identify microorganisms? DNA from an unknown organism is mixed with a DNA primer from a known organism for PCR. If the DNA is amplified in the PCR, the unknown organism is the same as the organism from which the primer was made.

3. Now that we have PCR, why do we use culture techniques? PCR can be used to classify and identify microbes, but it doesn't tell us about their physiology and what they are doing in the environment.

4. Identify some of the unusual habitats occupied by microorganisms. Some mentioned in this video are ice, boiling water, and the Earth's crust.
5. Which of the following is correct? Briefly explain why. b
   a. When an antibiotic is introduced, sensitive bacteria will develop resistance.
   b. When an antibiotic is introduced, resistant bacteria will survive and reproduce.

6. Why aren't viruses included in the tree of life? Viruses don't have ribosomes, so we can't construct a phylogeny that includes all other organisms and viruses.

EXERCISES

Concept Map 7.1

This map shows terms related to microbial diversity.

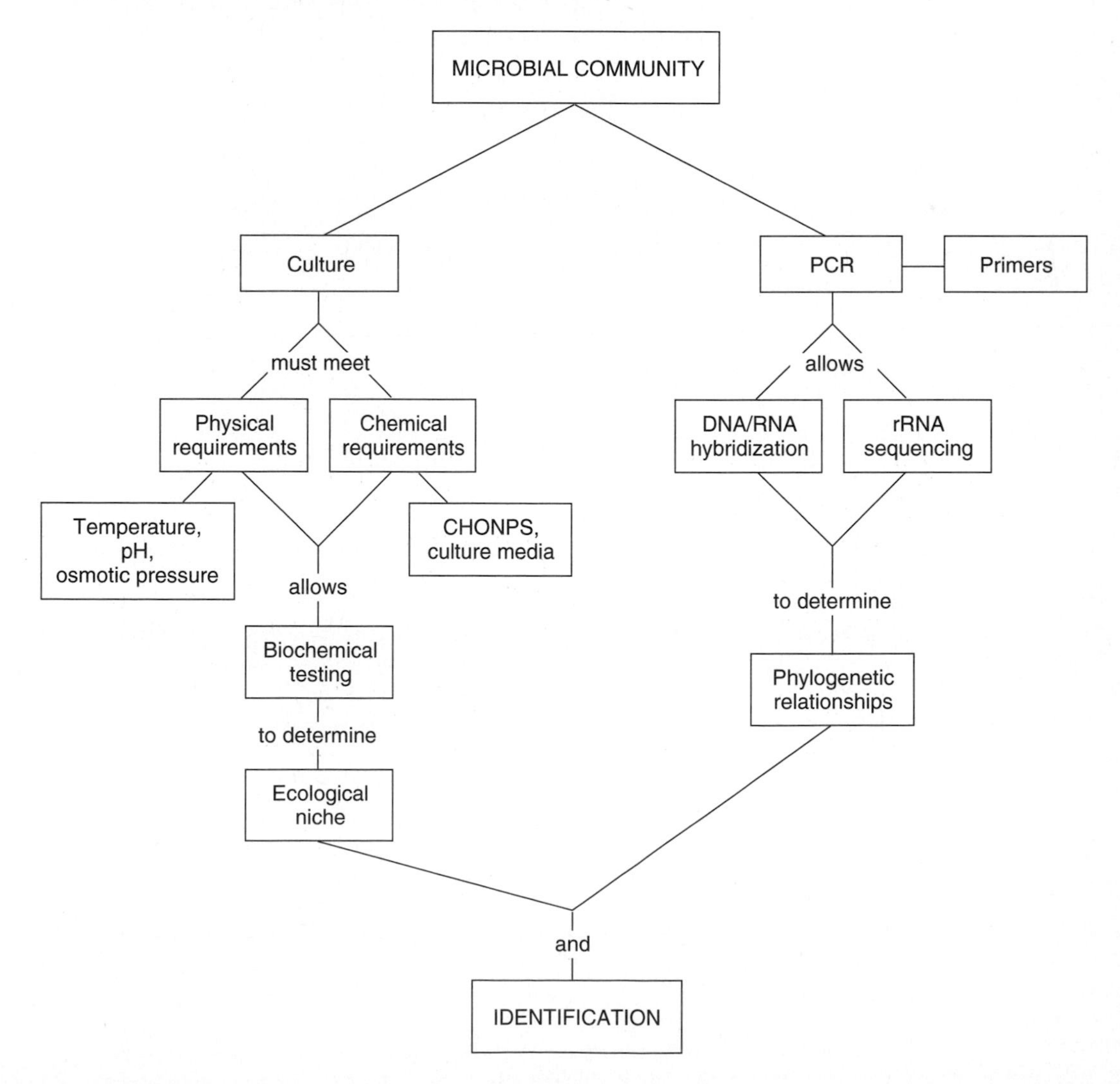

1 CHONPS refers to six chemical elements that all organisms need. What are the elements?
2. Where would you place an enrichment medium in this map?
3. Where would you place O_2?
4. Why are microbiologists using PCR instead of cultures to investigate soil bacteria?

Figure 7.1

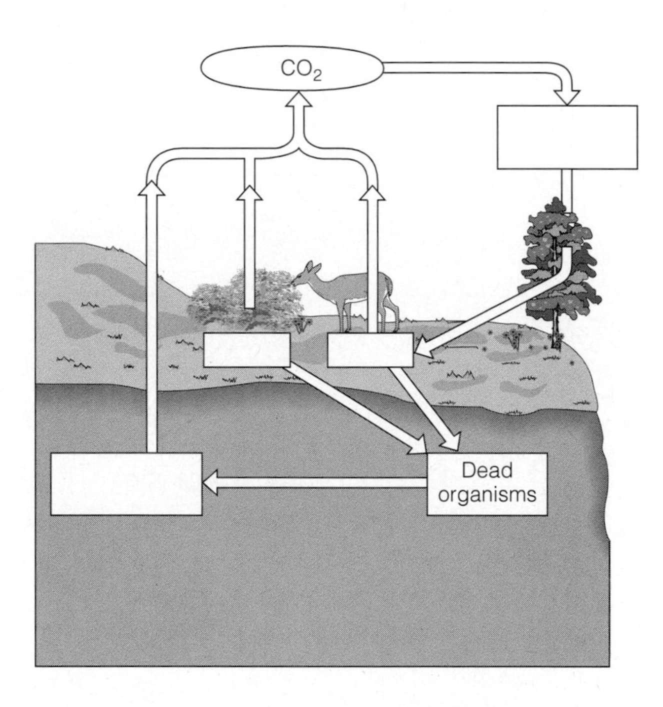

1. You learned the following terms in Unit 1 (Concept Map 1.2) and heard Tom Schmidt describe them in this video. Place the terms in their correct place in the figure.
   a. Consumers
   b. Producers
   c. Decomposers
   d. Cellulose-degrading bacteria
   e. Herbivores
   f. Carnivores
   g. Bacteria
   h. Fungi
   i. Photosynthesis
   j. A termite
2. Where would you place fossil fuels (e.g., coal, oil) in this carbon cycle?

3. Where would you place mollusk shells (i.e., $CaCO_3$) in this carbon cycle?

4. In the video, Woese asks: "Let's take all the bacteria in the world away. Do you know what's going to happen?"

Figure 7.2

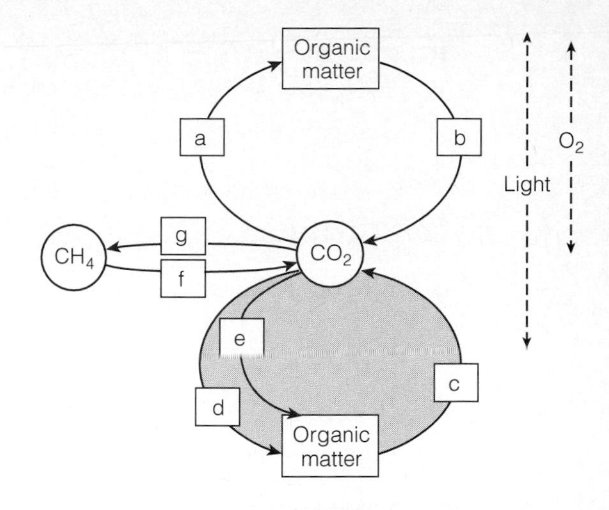

Place the following organisms in their correct positions in this carbon cycle. Review metabolic pathways in Unit 3 if necessary.

1. Cyanobacteria. Aerobic photoautotrophs; use cyclic photophosphorylation and the Calvin-Benson cycle. _____
2. *Gallionella*. Aerobic chemoautotrophs; use CO_2 for carbon; Fe^{2+} is the electron donor. _____
3. *Lactobacillus plantarum*. Anaerobic chemoheterotroph; strictly fermentative. _____
4. *Methanomicrobium*. Anaerobic, methane-producing Archaea; use CO_2 as the final electron acceptor. _____
5. *Methylomonas*. Aerobic bacteria; oxidize methane for carbon and energy. _____
6. *Pseudomonas aeruginosa*. Aerobic chemoheterotroph; uses a variety of carbohydrates and proteins and oxidative phosphorylation. _____
7. Purple sulfur bacterium. Anaerobic photoautotrophs; uses cyclic photophosphorylation; does not use water as an electron donor for photosynthesis; uses the Calvin-Benson cycle. _____
8. *Thermothrix thiopara.* A facultatively anaerobic, facultative chemoautotroph using CO_2 for carbon, H_2S for an electron donor, and NO_3^- as an electron acceptor. _____
9. At which step do producers belong? At which step do decomposers belong? _____
10. At which step do plants belong in this carbon cycle? At which step do animals belong? _____

Figure 7.3

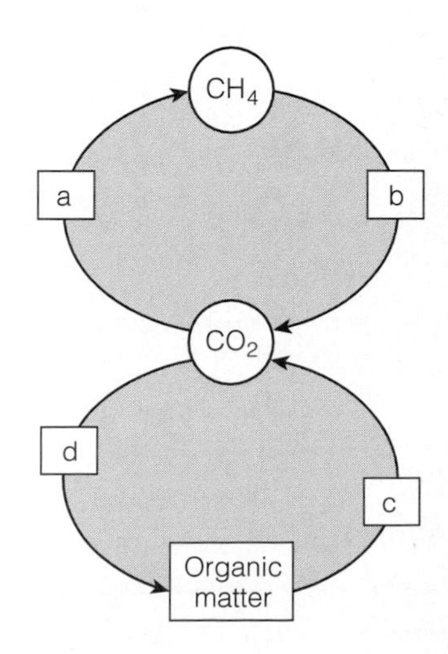

In Unit 3, you saw life deep under the ocean. Assume this is a diagram of the carbon cycle as it occurs in sediments at 2000 meters under the ocean. Remember, at that depth, there is no light. Place the following organisms in their proper places in the figure:

1. *Methanospirillum.* Use CO_2 as the final electron acceptor. _____
2. *Thiobacillus*. Use the Calvin-Benson cycle to fix CO_2. _____
3. An organism that oxidizes methane. _____
4. A strictly fermentative chemoheterotroph. _____

5. A phototrophic bacterium. ____
6. What are the primary producers in this ecosystem? What are they using for energy? Compare this to the ecosystems in Figures 7.1 and 7.2.

Definitions

Match the following statements to words from Key Terms and Concepts:

1. Bacteria found at >100°C in deep ocean vents. ____________________
2. Bacteria found in salt-evaporating ponds and in the Great Salt Lake. ____________________
3. Bacteria found at ~70°C in hot springs in Yellowstone National Park. ____________________
4. Enzyme that hydrolyzes H_2O_2 to produce O_2. ____________________
5. Enzyme that converts $O_2^- \cdot$ to H_2O_2. ____________________
6. Strictly fermentative bacteria that lack catalase and peroxidase, and can grow in the presence of air. ____________________
7. Bacteria with an electron transport chain that lack catalase and peroxidase, and cannot grow in the presence of air. ____________________

Classification Questions

The following names were in the video. Identify each as

a. Archaea
b. Bacteria
c. Eukarya

1. Arthropod ____
2. Chordate ____
3. Crenarchaeota ____
4. *Escherichia coli* ____
5. *Giardia* ____
6. Halobacteriaceae ____
7. *Methanosarcina* ____
8. *Nocardia* ____
9. Proteobacteria ____
10. *Thermotoga* ____
11. *Thermoplasma* ____

Habitat Diversity Questions

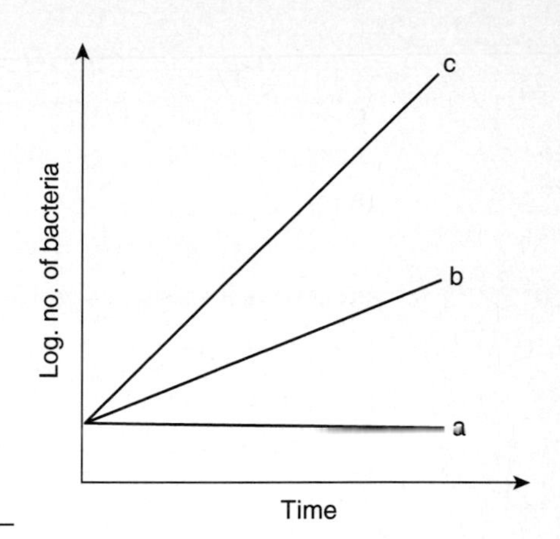

Use this graph to answer the following questions. Assume you are growing microorganisms and have provided an optimum environment, except for the variable in the question.

Which line best depicts each of the following?

1. An obligate anaerobe incubated aerobically. _____
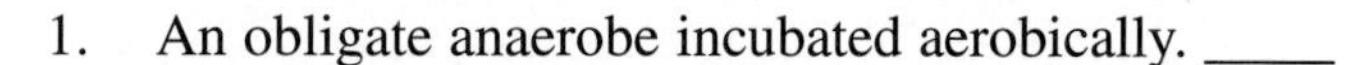
2. A facultative anaerobe incubated aerobically. _____
3. A facultative anaerobe incubated anaerobically. _____
4. An obligate halophile incubated in seawater. _____
5. A facultative halophile incubated in 7.5% salt. _____
6. A mesophile incubated at 50°C. _____
7. An extreme thermophile incubated at 37°C. _____
8. A mesophile in your refrigerator. _____

PCR Questions

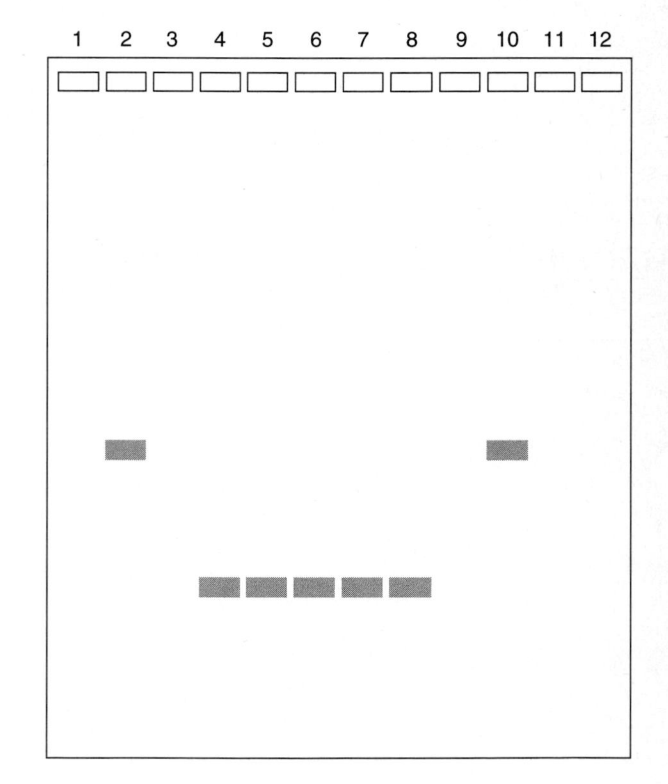

Schmidt's agarose gel is shown to the right. Lanes 2 and 10 are controls. Samples were loaded into wells 3–9.

1. Mark the + and – electrical poles.
2. Circle the DNA.
3. Which lanes contain amplified DNA? How can you tell?
4. The primer was for an rRNA gene. Why does this work if these microbes are unknown?

A Biotechnology Problem

Microbes from unusual environments are of interest to biotechology companies because they may produce new products or their enzymes may function under the extreme conditions often used in manufacturing. In Unit 4, you saw researchers looking for bacteria that produce new antibiotics. Identifying a new antibiotic is only part of the process of bringing a product to market. The antibiotic must be produced in sufficient quantities to test and then produce commerically.

Assume that you have isolated a bacterium that produces a new antibiotic. The bacterium grows well in Medium 1 but doesn't produce its antibiotic. Testing one component at a time, you determine that iron interferes with antibiotic production.

Medium	Modifications	Generation Time (min)	Antibiotic (mg/l)
1		27	0.02
2		300	0.20
1	Without Fe^{2+}	No growth	—
2	With Fe^{2+}	150	0.02

Design a procedure to maximize the biomass (total cells) and the yield of the antibiotic. Grow the cells to a high density in Medium 1, then transfer them to Medium 2 for antibiotic production.

Short-Answer Questions

1. Pasteur saw large oval cells in the sugar medium during wine preparation and small rods in wine as it turned to vinegar. This caused Pasteur to say that microorganisms could change shape dependent on their environment. Explain why you agree or disagree with his observations.

2. Assume you moistened a sterile swab in sterile water, swabbed a corner in your home, and inoculated a nutrient agar plate with that swab. Twenty-three colonies grew after incubation for 48 hr. at room temperature. Did all the bacteria from that corner of your home grow? Explain.

3. Most bacteria cannot metabolize agar, yet it is used in the majority of laboratory culture media. What is the purpose of the agar?

4. Where would you find a bacterium that is capable of metabolizing agar?

5. Of what value is PCR?

6. Why do culture and molecular (DNA) sampling results of the same habitat differ?

7. Look at Concept Map 7.1. Which is more important, identifying microorganisms or determining their ecological niche?

Hypothesis Testing

Hypothesis: *Thermotoga* is one of the closest living relatives of the first life on Earth.

Data/Observations: What evidence supports this hypothesis? Although classified as a bacterium, *Thermotoga* shares many genes with archaea and eukarya. It lives in a hot, sulfurous environment similar to the conditions when life first evolved on Earth.

Conclusions: From these data, do you accept or reject the hypothesis? Briefly explain.

Study Questions

***Microbiology: An Introduction,* Sixth Edition**

Pages	Review Questions	Multiple Choice Questions	Critical Thinking Questions	Clinical Applications Questions
151	7			
179–180	All	All	All	All
318–319	All	All	All	All
266			3	1
740	1–4			
760				3

CD Activities

Do the Chapter 6 and Chapter 9 quizzes.

Web Activities

1. Do the Chapter 6 and Chapter 9 quizzes.
2. Read about "Life on Mars" and answer the following questions:

a. Why look for microbial life instead of plants and animals? Large lifeforms would be visible in photographs taken from satellites or landing craft. Microbial life could be in the soil and rocks and, therefore, not seen.

b. What assumptions are made in attempts to look for life on Mars? That Martian life uses the same carbon sources as on Earth.

c. What is the control in the experiments? What is its purpose? The control is heated (sterilized) Martian soil inoculated into flasks of growth media. The control would detect any chemical reactions between the soil and media, so these changes would not be attributed to living organisms.

ANSWERS

Concept Map 7.1

1. Carbon, Hydrogen, Oxygen, Nitrogen, Phosphorus, Sulfur
2. Culture
3. Chemical requirements
4. Many bacteria and archaea haven't been cultured.

Figure 7.1

1. Note that (d) is inside (e). A termite could replace the deer.
2. Fossil fuels are sequestered organic molecules (dead plants and microorganisms) that are removed from the cycle.
3. $CaCO_3$ is mineralized inorganic carbon that is now removed from the cycle.
4. "All the life that we could look at, see with our naked eyes, all that life is going to disappear in short order. In other words, the microbial world is the basis upon which our whole ecosphere rests. And without it there is no multicellular life."

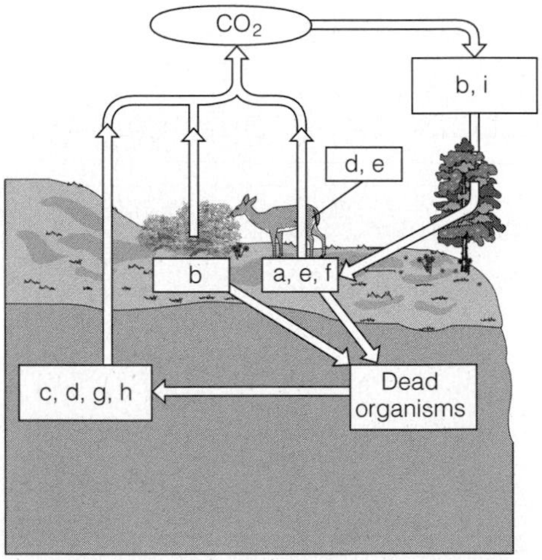

Figure 7.2

1. a
2. a
3. b, c
4. g (anaerobically)
5. f (aerobically)
6. b
7. e
8. d

9. Producers (a, e, d,); Decomposers (b, c)
10. Plants (a); Animals (b)

Figure 7.3

1. a
2. d
3. b
4. c
5. Not there
6. The primary producers are fixing CO_2 (d). They are using hydrogen and sulfur for energy (Units 3 and 6). In Figures 7.1 and 7.2, light is available for the primary producers.

Definitions

1. hyperthermophiles or extreme thermophiles
2. obligate halophiles
3. thermophiles
4. catalase
5. superoxide dismutase
6. aerotolerant anaerobes
7. obligate anaerobes

Classification Questions

1. c
2. c
3. a
4. b
5. c
6. a
7. a
8. b
9. b
10. Currently placed in b.
11. a

Habitat Diversity Questions

1. a
2. c
3. b
4. a
5. b
6. a or b
7. a
8. b

PCR Questions

1. The – pole is at the start (wells).

2–3. Lanes 4–8 because the DNA is visible. The primer was for a highly conserved region of the rRNA gene, so all organisms will have it.

4. The primer is for a highly conserved region of rRNA that all cells should have.

Short-Answer Questions

1. The shapes of microbes are generally fixed genetically. Pasteur was not aware that he was looking at a microbial succession. The yeast changed the environment and could no longer grow, but the alcohol provided a carbon source for *Acetobacter*.
2. Probably not. Only organisms that could grow on the carbon and energy source provided and at that temperature, within 48 hr., grew.
3. The agar is used as a solidifying agent.
4. Agar-digesting bacteria can be found decomposing macroscopic algae.
5. PCR is used to make large quantities of DNA for use in DNA fingerprinting or genetic engineering (Unit 5).
6. (1) Culture methods may not provide the correct nutrients for all organisms. (2) During enrichment, some microbes may be killed by metabolic products of other organisms. (3) Some organisms may need specific products of other organisms that did not grow.
7. The answer depends on what you are trying to do. Some researchers are determining how microbes degrade toxic wastes or provide food in a food chain. Other researchers are determining how all organisms are related. In either case, it is helpful to be able to have an identifier, a unique name or even a number, when referring to a microbe.

Unit 8

MICROBIAL ECOLOGY

The role of the infinitely small appears to me infinitely large.
—LOUIS PASTEUR, QUOTED BY SERGEI WINOGRADSKY, 1896

LEARNING OBJECTIVES

	1.	Define biogeochemical cycle.	p. 716
★	2.	Outline the carbon cycle, and explain the roles of microorganisms in this cycle. (Also in Unit 7.)	p. 716
★	3.	Outline the nitrogen cycle, and explain the roles of microorganisms in this cycle.	p. 717
★	4.	Define ammonification, nitrification, denitrification, and nitrogen fixation.	p. 717
	5.	Outline the sulfur cycle, and explain the roles of microorganisms in this cycle.	p. 721
	6.	Compare and contrast the carbon cycle and the phosphorus cycle.	p. 723
★	7.	Define bioremediation. (Also in Unit 10.)	p. 724
★	8.	Give two examples of the use of bacteria to remove pollutants. (Also in Unit 10.)	p. 724
	9.	Describe the freshwater and seawater habitats of microorganisms.	p. 725
★	10.	Define BOD.	p. 731

READING

Chapter 27 (pp. 715, 717–727, 732), pp. 35, 301, 751.

SUGGESTED LABS FROM JOHNSON AND CASE, *LABORATORY EXPERIMENTS IN MICROBIOLOGY*, FIFTH EDITION

Exercise 55: Microbes in Soil: The Nitrogen and Sulfur Cycles
Exercise 56: Microbes in Soil: Bioremediation

KEY TERMS AND CONCEPTS

acid precipitation p. 723
ammonification p. 717
bacteroids p. 720
biochemical oxygen demand p. 732
biogeochemical cycles p. 716
bioluminescence p. 726
bioremediation p. 724
carbon cycle p. 716
deamination p. 717
denitrification p. 719
dissimilation p. 723
global warming p. 717
heterocysts p. 720
infection thread p. 720
lichens p. 720
nitrification p. 718
nitrogen cycle p. 717
nitrogen fixation p. 720
phosphorus cycle p. 723
root nodules p. 720
sulfur cycle p. 721

INTRODUCTION

Be sure to read the Study Outlines for Chapter 27, pp. 738–740, and Chapter 28, p. 758.

Microbes, especially bacteria, live in the most widely varied habitats on Earth. The ability of bacteria to live in so many habitats is due to their ability to grow under different physical conditions. Additionally, bacteria can occupy a variety of niches due to their metabolic diversity—that is, bacteria can use a variety of carbon and energy sources.

Bacteria are essential to each ecosystem on Earth. As you saw in Unit 1, one of their functions is to decompose organic material to recycle it for plant and algal use. This is usually referred to as the carbon cycle. Carbon is not the only element that must be recycled by bacteria; many chemical elements are metabolized and oxidized or reduced by organisms in biogeochemical cycles. The amount of each element on the Earth is fixed, and each organism can usually only use an element in a specific form. For example, plants use sulfur in the form of sulfate (SO_4^{2-}) which they reduce to S^{2-} for the sulfhydral groups (—SH) of proteins. The decomposers release the sulfur that is locked up in proteins and make it available for plants. As Pasteur observed, the metabolic activities of bacteria are vital for continuation of these biogeochemical cycles.

Our knowledge of biogeochemical cycles is increasing, and we are learning to use microbes to clean up the wastes produced by a large technological society. One of the first such uses was the sewage treatment plant you saw in Unit 3. The first municipal sewage treatment in the late 1940s, primary treatment, was designed to improve public health by killing pathogenic microbes. In the 1960s, people became aware of the need to remove organic matter from wastewater to improve or maintain the ecology of waterways. The use of microbes to degrade or detoxify pollutants is called bioremediation.

Although the activities of most microbes are beneficial to the ecosystem, a few microbes do cause food spoilage or disease. The control of unwanted microbes is the subject of Unit 9.

VIDEO: MICROBIAL ECOLOGY

Terms

The following new terms are introduced in this video:

Biosphere I	estuary
Biosphere II	herbivore
ecosphere	*Spartina*

Preview of Video Program

This video program begins at a lake at Mt. St. Helens to explain how microbial growth changes the environment. The lake began as a mudflow from the Mt. St. Helens eruption in 1980. There were no frogs or fish, and if any came, they would not have been able to grow in the turbid, sulfurous water. Some microbes could grow, however, and as they grew, they changed the environment. Eventually, other microbes and finally plants and animals could also grow there.

The video shows you Biosphere II, which was an attempt to make a self-contained ecosystem to gather information necessary for humans to inhabit other planets. Biosphere II included plants to fix CO_2 and produce O_2 so the animals could live. Bacteria present in the soil and compost would recycle CO_2 and other inorganic compounds for the plants. An unforeseen imbalance due to the buildup of organic matter caused a severe problem for Biosphere II.

Salt marsh ecology is covered next. The salt marsh is being studied to find out more about the activities of microbes in the ecosystem. Hans Pearl of the University of North Carolina explains a salt-marsh ecosystem in which the primary producer is *Spartina*. In Unit 1, you probably placed bacteria at the end of the food chain, to start the cycle again. But Pearl shows that some heterotrophic bacteria occupy a different niche in the food chain; they are not recycling nutrients back to plants.

Not all microbial communities are in the external environment. Many microbes live as symbionts inside another organism. Like the termites seen in Unit 7, cows eat cellulose but cannot digest it. Cows have a specialized organ in their digestive system, called a rumen, that houses a diverse bacterial population. You will learn more about the rumen in Unit 10.

Recall the sewage treatment plant you saw in Unit 3. Primary treatment (sedimentation) leaves organic matter in the water that is discharged to a river or ocean. Bacteria will degrade this organic matter using oxygen, creating a biochemical oxygen demand, or BOD. Secondary sewage treatment (aerobic digestion) is designed to move this organic matter before the water is discharged. Sewage treatment is an example of bioremediation. Another example of bioremediation is shown at the Columbia County (Georgia) Land Fill. Modern societies accumulate a great deal of waste that is kept sealed in landfills. Terry Hazen has developed a process to increase the rate of decomposition of this solid waste.

Video Questions

1. Differentiate between light and dark clouds. How does this affect the Earth's temperature? How do microbes contribute to the cloud color? Marine algae produce a sulfur compound, which escapes from the ocean into the atmosphere. In the atmosphere, the sulfur particles cause moisture in the air to form cloud droplets. With many droplets, we see a very white cloud, with just a few, it looks dark. The degree of whiteness affects how much of the sun's light and heat are reflected from Earth back into space, one of the things that controls the Earth's temperature.

2. Why did Biosphere II fail? Use BOD in your answer. The organic matter from crops and animals caused a BOD. Bacteria decomposing the organic matter used the O_2 in the atmosphere faster than it was being created.

3. What does Hazen add to the landfill to promote microbial growth? Air and water.

4. What causes the high temperature in Hazen's landfill experiment? Metabolic activity of microbes.

EXERCISES

Concept Map 8.1

This map shows terms related to ecosystems.

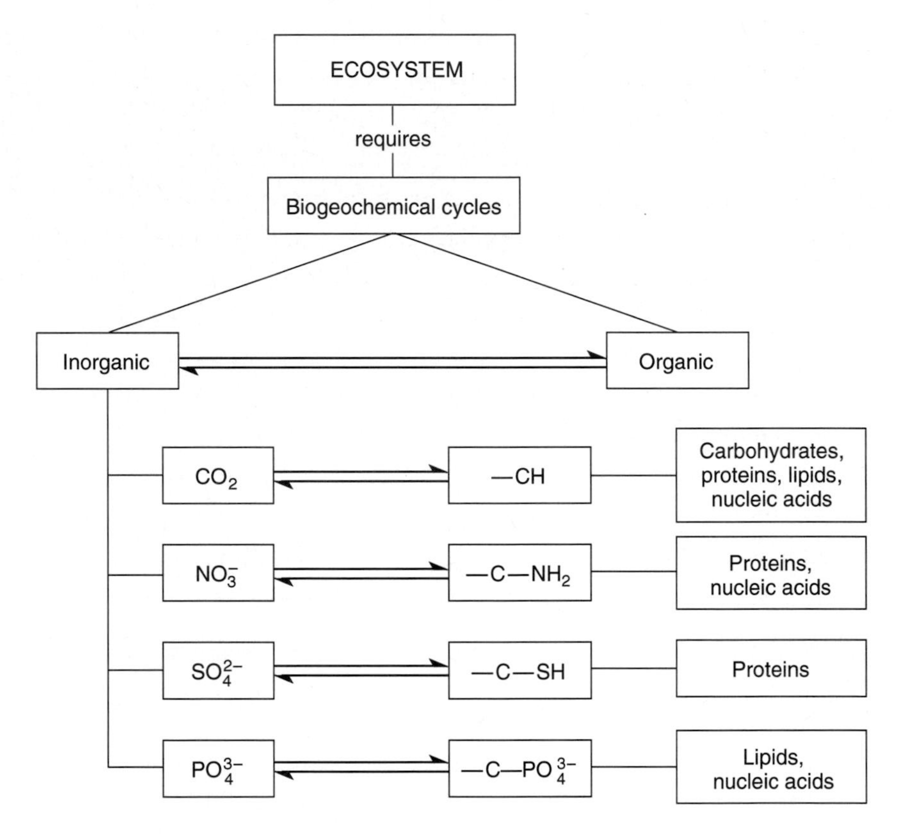

Place the following terms in their proper places:

1. *Spartina*
2. Algae
3. Decomposers
4. Plants
5. Animals
6. Calvin-Benson cycle

Figure 8.1

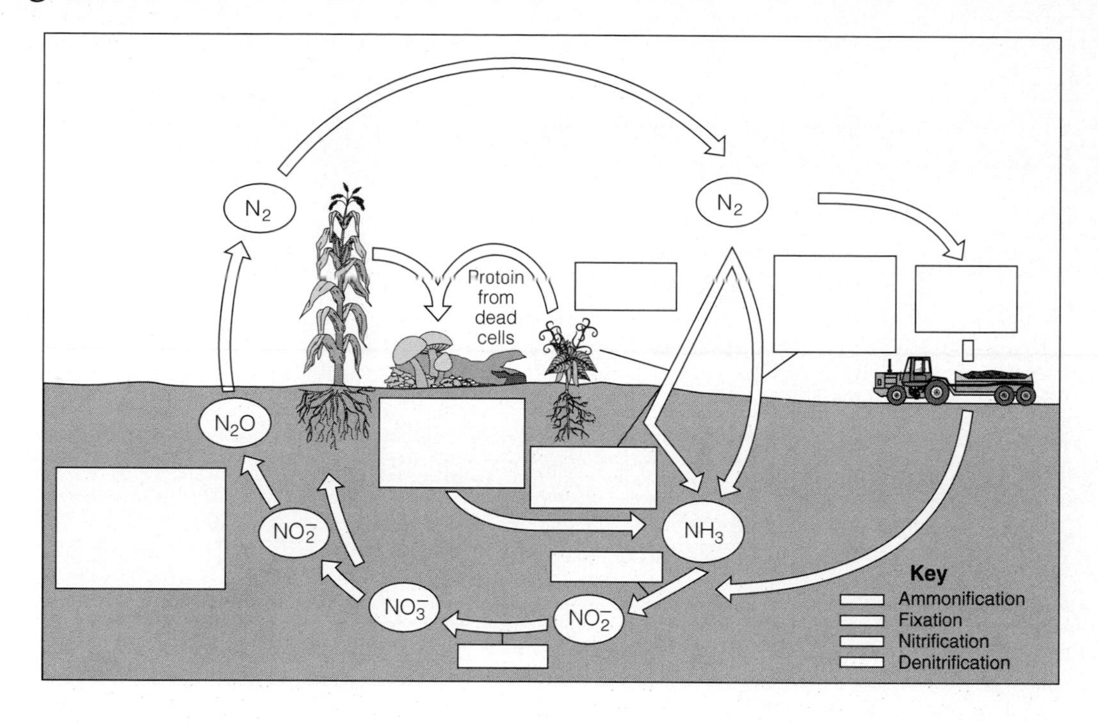

1. Color the arrows and key to indicate where the following processes are occurring:
   a. Ammonification
   b. Denitrification
   c. Nitrification
   d. Nitrogen fixation
2. Which process(s) is (are) anaerobic respiration?

3. Where are the chemicals containing the —NH_2 group?

4. Place the following bacteria in their correct locations:
   a. Chemoautotrophs that use nitrogen atoms for their energy source
   b. Chemoheterotrophs that hydrolyze protein
   c. Cyanobacteria
   d. *Nitrobacter*
   e. *Nitrosomonas*
   f. *Paracoccus denitrificans*
   g. *Rhizobium*
5. Why is nitrogen needed by organisms?

6. Which group of microorganisms compete with plants for nitrogen?

7. To what do you attribute the smell of ammonia in a baby's diapers?

8. Is this an oxidation or reduction reaction: $NO_2^- \rightarrow NO_3^-$? How can you tell?

Figure 8.2

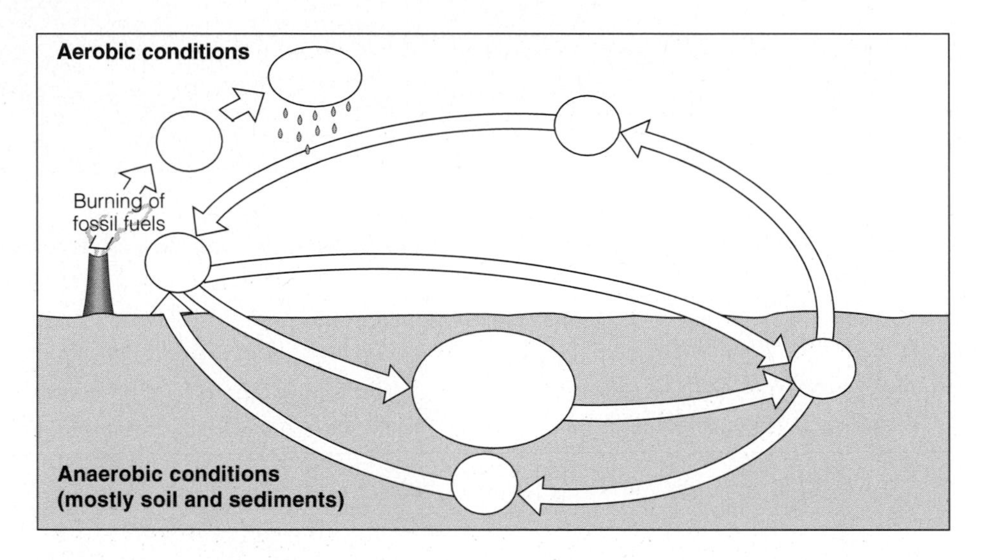

1. Place the following in the correct circles:
   a. —SH
   b. H_2S
   c. H_2SO_3
   d. S^0
   e. SO_2
   f. SO_4^{2-}
2. Color the arrows to indicate where the following processes are occurring:
   a. Assimilation of sulfur
   b. Dissimilation of sulfur-containing compounds
   c. Formation of acid precipitation
   d. Oxidation of sulfur
   e. Photosynthesis
   f. Reduction of sulfate ion

3. Which process(es) use sulfur as an energy source?

4. Why is sulfur needed by living organisms?

5. In which process(es) is sulfur a by-product and not the main reactant?

Definitions

Match the following statements to words from Key Terms and Concepts:

1. Release of sulfur from organic molecules. ____________________
2. A measure of biologically degradable organic matter in water. ____________________
3. Alternating oxidation and reduction of chemical elements. ____________________
4. A symbiosis between a cyanobacterium and a fungus. ____________________
5. The process by which luciferase releases a photon of light. ____________________

Matching Questions

Match the following choices to the statements below:

a. Herons
b. Fish
c. *Spartina*
d. Crab larvae
e. Bacteria

1. Producers. ____
2. Decomposers. ____
3. Food for herons. ____
4. Food for crab larvae. ____
5. Fix nitrogen. ____
6. In the biofilm on the *Spartina*. ____

A Bioremediation Problem

Mercury is a common pollutant due to leakage from dumps, paper manufacture, and gold mining. Diagram a mercury cycle. Can bioremediation solve the mercury pollution problem? The necessary information is on p. 35 in the textbook.

Short-Answer Questions

1. Occasionally a newspaper headline will state that a "sewage spill kills fish." The fish deaths often occur several days after the sewage spill. If the sewage is not toxic, what causes the fish to die?

2. Why are lichens and legumes early colonizers on newly exposed rocks?

3. What is the significance of dumping organic matter from a primary sewage treatment plant, for example, into a body of water?

4. What is the significance of high nitrate ion and phosphate ion concentrations in a body of water?

5. Your friend carefully bleached and scrubbed all the rocks in his aquarium then put them back into the tank with clean water. Within a few days, his tropical fish began dying. His chlorine test showed that no chlorine was present in the aquarium. What do you advise him to do?

Hypothesis Testing

Hypothesis: Microbes contribute to global warming.

Data/Observations: What evidence supports this hypothesis? Bacteria produce methane, and algae produce sulfur compounds and chlorofluorocarbons. These three chemicals are "greenhouse gases." Ask the students whether they can find a microbial solution to this problem.

Conclusion: From these data, do you accept or reject the hypothesis? Briefly explain.

Study Questions

Microbiology: An Introduction, Sixth Edition				
Pages	**Review Questions**	**Multiple Choice Questions**	**Critical Thinking Questions**	**Clinical Applications Questions**
740–741	5–7, 12, 14–15		1, 2	1, 2

CD Activity

Do the Chapter 27 quiz.

Web Activity

Do the Chapter 27 quiz.

Outdoor Activity

Look in your lawn or a vacant lot for a legume, a plant in the pea family. Clover, vetch, and lupines are examples. How will you recognize a legume? Students may need a description of the flowers and to be told that identification cannot be based on leaves alone. The flowers have five petals. The lateral petals form "wings," and the bottom two petals come together to form a keel.

Dig up the plant's roots. Compare the roots to a plant in another family. What color are the root nodules? pinkish Sketch the root nodules. What would you see if you examined the nodules microscopically? Gram-negative bacteria.

ANSWERS

Concept Map 8.1

1. Assimilate inorganic ions and molecules into organic compounds.
2. Assimilate inorganic ions and molecules into organic compounds.

3. Dissimilate organic compounds into inorganic ions and molecules.
4. Assimilate inorganic ions and molecules into organic compounds.
5. Animals use the organic compounds made by plants and rearrange them into their own bodies. Energy-producing reactions release CO_2.
6. The Calvin-Benson cycle is used by *Spartina* and other plants and algae to fix CO_2 into organic compounds (—CH).

Figure 8.1

1. See Figure 27.3 (p. 718) in your textbook.
2. Denitrification
3. In organisms
4. a. Nitrification
   b. Ammonification
   c. Nitrogen fixation and assimilation
   d. Nitrification
   e. Nitrification
   f. Denitrification
   g. Nitrogen fixation and assimilation
5. To synthesize proteins and nucleic acids
6. Denitrifiers
7. Breakdown of urea to ammonia
8. It's an oxidation reaction because the nitrogen lost 2 electrons: $N^{3+} \rightarrow N^{5+}$.

Figure 8.2

1. See Figure 27.6 (p. 722) in your textbook.
2. a, b, and c: See Figure 27.6 (p. 722) in your textbook.
   d. $H_2S \rightarrow SO_4^{2-}$
   e. Phototrophic bacteria use H_2S or S^0 as electron donors.
   f. $SO_4^{2-} \rightarrow$ —SH or H_2S
3. $H_2S \rightarrow SO_4^{2-}$ in the aerobic pathways (e.g., *Thiobacillus*)
4. For protein synthesis
5. $H_2S \rightarrow S^0$; phototrophic bacteria using H_2S as an electron donor to fix CO_2.

Definitions

1. dissimilation
2. biochemical oxygen demand
3. biogeochemical cycles
4. lichen
5. bioluminescence

Matching Questions

1. c
2. e
3. b

4. e
5. e
6. e

Short-Answer Questions

1. The sewage increased the BOD in the water. Bacteria used the dissolved oxygen (O_2) in the water to degrade the organic matter in the sewage and deprived the fish of oxygen.
2. They are capable of fixing nitrogen, which would be lacking in rocks.
3. The organic matter will increase the BOD in the water.
4. These nutrients will promote algal growth. Excess algae will lead to an increase in organic matter and an increased BOD.
5. As fish metabolize proteins, they excrete ammonium ions from their gills. Ammonium was probably accumulating to toxic levels in the aquarium. He needs to replace the nitrifying bacteria that formed the biofilm on the rocks. The easiest way would be to get some rocks from an established aquarium.

Unit **9**

MICROBIAL CONTROL

We must seriously consider the possibility that . . . resistant strains can arise, the important question [then is], what therapeutic measures can be taken in such a case?
—Paul Ehrlich, 1909

Learning Objectives

★	1.	Define the following key terms related to microbial control: sterilization, disinfection, antisepsis, germicide, bacteriostasis, asepsis, degerming, and sanitation.	p. 181
	2.	Describe the patterns of microbial death caused by treatments with microbial control agents.	p. 182
	3.	Describe the effects of microbial control agents on cellular structures.	p. 183
	4.	Compare the effectiveness of moist heat (boiling, autoclaving, pasteurization) and dry heat.	p. 184
	5.	Describe how filtration, low temperatures, desiccation, and osmotic pressure suppress microbial growth.	p. 184
	6.	Explain how radiation kills cells.	p. 184
	7.	Explain how microbial growth is affected by the type of microbe and the environmental conditions.	p. 191
	8.	List the factors related to effective disinfection.	p. 191
	9.	Interpret the results of use-dilution and filter paper tests.	p. 192
	10.	Identify the methods of action and preferred uses of chemical disinfectants.	p. 193
	11.	Differentiate between halogens used as antiseptics and as disinfectants.	p. 193
	12.	Identify the appropriate uses for surface-active agents.	p. 193
	13.	List the advantages of glutaraldehyde over other chemical disinfectants.	p. 193
	14.	Identify the method of sterilizing plastic labware.	p. 193
	15.	Define the contributions of Paul Ehrlich and Alexander Fleming to chemotherapy.	p. 531
	16.	Identify the microbes that produce most antibiotics.	p. 531
★	17.	Describe the problems of chemotherapy for viral, fungal, protozoan, and helminthic infections.	p. 532
	18.	Define the following terms: spectrum of activity, broad-spectrum antibiotic, superinfection.	p. 532
	19.	Identify five modes of action of antimicrobial drugs.	p. 533
	20.	Explain why the drugs described in this section are specific for bacteria.	p. 536
	21.	List the advantages of each of the following over penicillin: semisynthetic penicillins, cephalosporins, and vancomycin.	p. 536
	22.	Explain why INH and ethambutol are antimycobacterial agents.	p. 536
	23.	Describe how each of the following inhibits protein synthesis: aminoglycosides, tetracyclines, chloramphenicol, macrolides.	p. 541

24. Describe how rifamycins and quinolones kill bacteria. p. 542
25. Describe how sulfa drugs inhibit microbial growth. p. 543
26. Explain the modes of action of currently used antifungal drugs. p. 543
27. Explain the modes of action of currently used antiviral drugs. p. 545
28. Explain the modes of action of currently used antiprotozoan and antihelminthic drugs. p. 547
★ 29. Describe two tests for microbial susceptibility to chemotherapeutic agents. p. 549
30. Describe the mechanisms of drug resistance. p. 550
31. Compare and contrast synergism and antagonism. p. 551
32. Describe thermophilic anaerobic spoilage and flat sour spoilage by mesophilic bacteria. p. 742
★ 33. Compare and contrast food preservation by industrial food canning, aseptic packaging, and radiation. p. 544

READING

Chapters 7 and 20 and pp. 411, 742–745.

SUGGESTED LABS FROM JOHNSON AND CASE, *LABORATORY EXPERIMENTS IN MICROBIOLOGY,* FIFTH EDITION

Exercises 22–26: Control of Microbial Growth

KEY TERMS AND CONCEPTS

antagonism p. 552
antibiotics p. 531
antimicrobial drugs p. 531
antisepsis p. 182
asepsis p. 182
aseptic packaging p. 744
autoclave p. 185
bactericidal p. 533
bacteriostasis p. 182
bacteriostatic p. 533
broad-spectrum antibiotics p. 532
broth dilution test p. 549
chemotherapeutic agents p. 531
chemotherapy p. 531
commercial sterilization p. 181
decimal reduction time (DRT, D value) p. 184
degerming p. 182
disinfection p. 182
disk-diffusion method p. 549
equivalent treatments p. 187
E test p. 549
flaming p. 187
flat sour spoilage p. 743
germicide p. 182
high-efficiency particulate air (HEPA) filters p. 187
high-temperature short-time (HTST) pasteurization p. 187
hot-air sterilization p. 187
ionizing radiation p. 189
membrane filters p. 187
microwaves p. 191
minimal bactericidal concentration (MBC) p. 549
minimal inhibitory concentration (MIC) p. 549
nonionizing radiation p. 189
oligodynamic action p. 195
resistance (R) factors p. 551
sanitization p. 182
selective toxicity p. 531
sepsis p. 182
spectrum of microbial activity p. 532
sterilization p. 181
superinfection p. 533
synergism pp. 543, 551
synthetic drugs p. 531
thermal death point (TDP) p. 184
thermal death time (TDT) p. 184
thermophilic anaerobic spoilage p. 743
ultra-high-temperature (UHT) treatments p. 187
use-dilution test p. 192
zone of inhibition p. 549

Introduction

Be sure to read the Study Outlines for Chapter 7, pp. 200–203, and Chapter 20, pp. 553–554.

Health care workers and microbiologists use aseptic techniques to eliminate unwanted microorganisms. The laboratory procedures shown in Units 3, 4, and 9 are standard practices that employ Biosafety Level 1 or 2 controls. You'll see Biosafety Level 4 in Unit 12. (The Biosafety Levels are described in Appendix C of this book.) Sterilization, the removal of all microorganisms, is essential in some environments, such as in microbiology culture media, surgical equipment, or pharmaceutical manufacturing. More often, disinfection is sufficient and a more practical microbial control.

Physical methods of control are often used to achieve sterilization. The most commonly used method is steam under pressure, in an autoclave. The autoclave is fast and inexpensive to run and will sterilize objects and solutions; however, materials placed in the autoclave must not be affected by heat and moisture. Many materials used in microbiology and medicine are plastic, and certain solutions, such as antibiotics, could be denatured by heat. Then an alternate method of sterilization has to be chosen.

Chemicals can be used as disinfectants to decontaminate environmental surfaces such as laboratory work benches or home bathrooms. Most chemicals cannot sterilize. Antiseptics are chemicals used to decontaminate skin. Disinfectants can be more toxic than antiseptics because they aren't designed for use on skin.

Chemical disinfectants and antiseptics are not uniformly effective against all microorganisms. Gram-positive bacteria are generally more resistant than gram-negative bacteria, and mycobacteria are especially resistant because of their waxy cell wall. *Pseudomonas* bacteria are often resistant because they can degrade a variety of unusual compounds including many disinfectants. Enveloped viruses are more resistant than nonenveloped viruses. Bacterial endospores and protozoan cysts are more resistant than vegetative cells.

Name a disease caused by the following:

Mycobacterium: ____________________

Pseudomonas: ____________________

An enveloped virus: ____________________

A nonenveloped virus: ____________________

Most people are familiar with chemotherapy for cancer. Remember that chemotherapy means chemical treatment; the term doesn't specify the target of the treatment. Antimicrobial chemotherapeutic agents are used to kill or inhibit a microorganism growing in a body, causing disease. Chemotherapeutic agents differ from antiseptics and disinfectants because they work inside the body, so they must not be toxic to the host's cells. Antibiotics are antimicrobial chemotherapeutics produced by bacteria and fungi. Synthetic drugs are antimicrobial chemotherapeutics that are chemically synthesized in a lab. Semisynthetic antibiotics combine both techniques. Semisynthetic penicillins use the ß-lactam ring produced by *Penicillium*, then a chemical group is added to it to make a more effective antibiotic than natural penicillin.

Prior to the discovery and use of penicillin, people often died from bacterial infections. As recently as the 1930s, diseases that were not considered treatable, such as streptococcal pneumonia, were fatal. The discovery of an antibiotic-producing *Bacillus* in the 1930s spurred microbiologists to look for other common microorganisms that produce antibiotics.

In 1952, Selman Waksman received the Nobel prize for his discovery of streptomycin-producing *Streptomyces* bacteria. Since then, hundreds of antibiotic-producing bacteria have been discovered. As this unit's opening quote attests, drug resistance was seen as early as 1909. In Unit 4, you saw Julian Davies looking for new antibiotics in soil bacteria. The disk-diffusion method he used is a standard test for determining susceptibility to antibiotics. In Unit 9, you will see the disk-diffusion method used in the clinical environment.

Antibiotics are used to treat bacterial diseases; viruses and eukaryotic pathogens are not affected by antibiotics. Drugs used to treat viral, fungal, protozoan, and helminthic diseases are more difficult to develop because the target cells are eukaryotic, just like the human host. Recall from Unit 2 that viruses do not have cells but they reproduce by using the host cell's ribosomes, energy, and enzymes. Most of the antiviral drugs are nucleoside analogs; you saw in Unit 4 that these chemicals can stop DNA synthesis or cause mutations. A newer group of antiviral drugs, such as protease inhibitors, work by competitive inhibition against a viral enzyme. Enzymes were discussed in Unit 3.

There is a whole range of relationships between macroorganisms and microorganisms—from helpful to harmful. These relationships will be discussed in Units 10–12.

VIDEO: MICROBIAL CONTROL

Term

The following new term is introduced in this video:
selective medium

Preview of Video Program

In Unit 3, you saw microorganisms used to make food. In this video program you will see the need to kill or inhibit microorganisms in certain controlled environments such as the space program, hospitals, and the canning industry. Food production, storage, and handling are discussed. As Jim Leinfelder walks through a supermarket, we see that foods are frequently preserved by canning or refrigeration. You can read more about sterilizing spacecraft on p. 199 in your textbook.

Many examples of aseptic techniques, from handwashing to a surgical scrub, are shown in a hospital. Hospital infection control is described by Paul Lewis. You can read about the hospital infection control officer on p. 411 in your textbook. As you learned in Unit 4, natural selection is always at work, and the microorganisms that are resistant to the chemotherapeutic drugs we use survive and reproduce. This antibiotic resistance is often carried on plasmids (Unit 5). Lewis describes the formation of a colony from a single cell, as you learned in Unit 3. Lewis demonstrates another lab technique called the disk-diffusion test, which is used to look for antibiotic-resistant bacteria in the hospital. He uses a selective medium containing nutrients plus an antibiotic to grow bacteria that are resistant to the antibiotic.

Today, people are susceptible to far fewer bacterial infections due to the use of vaccines and antibiotics. (Vaccines are the subject of Unit 11.) Yet infectious diseases are the world's biggest killers. Germaine Hanquet of Doctors Without Borders explains the problems in developing countries. Worldwide, 50,000 people a day die of infectious diseases such as typhoid, tuberculosis, cholera, yellow fever, and malaria.

Video Questions

1. Write definitions of the following terms as you watch the video: (As shown on slates in the video.)

   a. Disinfection. The destruction of pathogenic organisms on inanimate objects.

   b. Antisepsis. The destruction of pathogenic organisms on a living object.

   c. Sterilization. The destruction of all forms of microbial life.

2. Describe the four levels of precautions in the clinical infection control program and explain when each is used.

   a. Level 1. Handwashing. Before and after any patient contact.

   b. Level 2. Contact precautions; gloves and gown. For communicable diseases that are not transmitted by the respiratory route.

   c. Level 3. Droplet precautions; gloves, gown, and mask. For respiratory infections.

   d. Level 4. Airborne precautions; patient isolated in a low-pressure room, staff wear high-efficiency filter masks in addition to gloves and gown. For highly contagious diseases.

3. What precautions should be taken in the home kitchen while preparing food? Use a different knife and cutting board for items that are not cooked before eating; wash hands before handling items that are not cooked before eating.

4. What reasons are given by Hanquet for the prevalence of infectious diseases? Contaminated water, lack of education, and poor hygiene. This problem will be addressed again in Unit 12.

5. Use your textbook to complete the following table:

Disease	Causative Agent	Method of Transmission	Treatment	Prevention
Typhoid	*Salmonella enterica*	Fecal-oral	Fluid and salts	Cooking food; treating water
Tuberculosis	*Mycobacterium tuberculosis*	Respiratory	Antimycobacterial drugs	Vaccine
Cholera	*Vibrio cholerae*	Fecal-oral	Fluid and salts	Cooking food; treating water
Yellow fever	An arbovirus	*Aedes aegyptii* mosquito	None	Vaccine
Malaria	*Plasmodium* spp.	*Anopheles* mosquito	Antimalarial drugs	Mosquito nets and repellents; mosquito control

EXERCISES

Concept Map 9.1

This map shows terms related to controlling microbial growth.

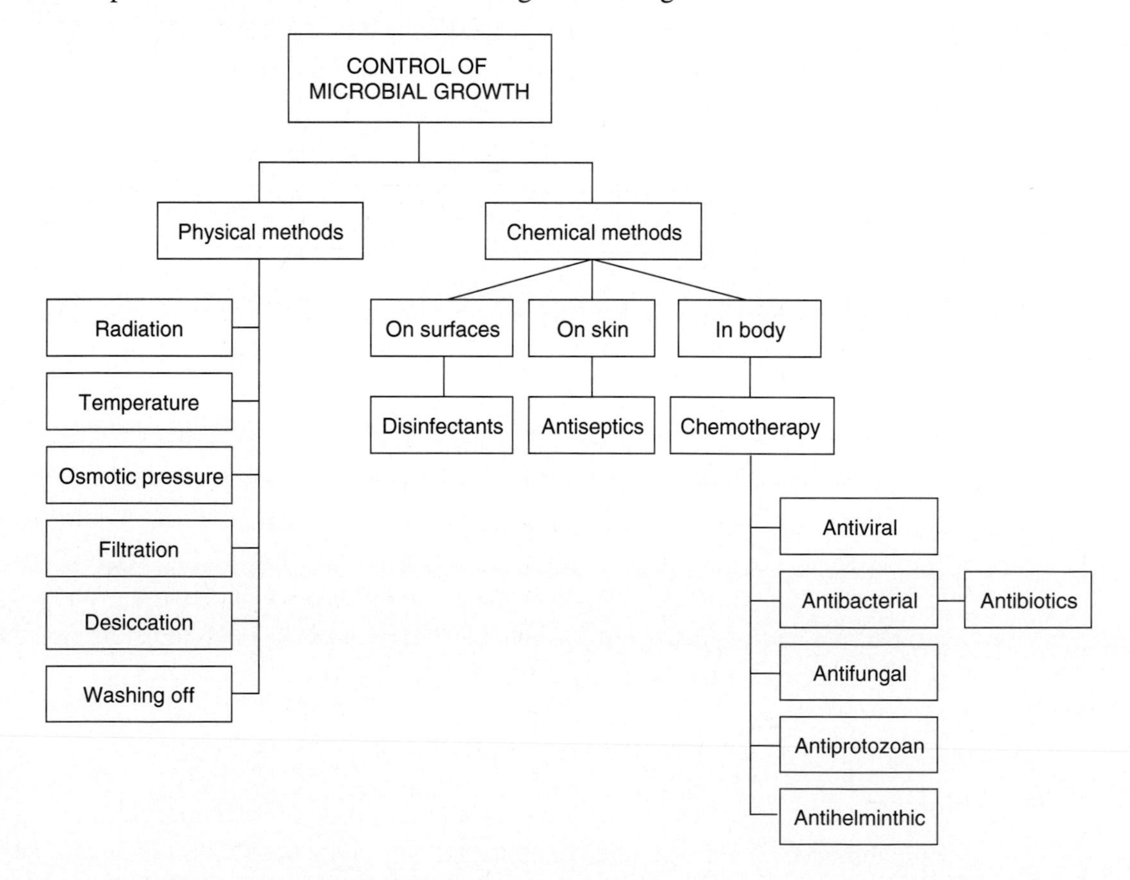

1. Fill in the types of radiation and temperatures used.
2. Which methods shown in the map can sterilize?

3. Give an example of an application of each of the physical methods of control.
4. Which category would be used to treat the following:
   a. A pimple?
   b. *Trichophyton?*
   c. Influenza?
   d. *Enterobius vermicularis?*
   e. Streptoccocal pharyngitis?
5. Why doesn't one drug work for all of the examples in question 4?

Figure 9.1

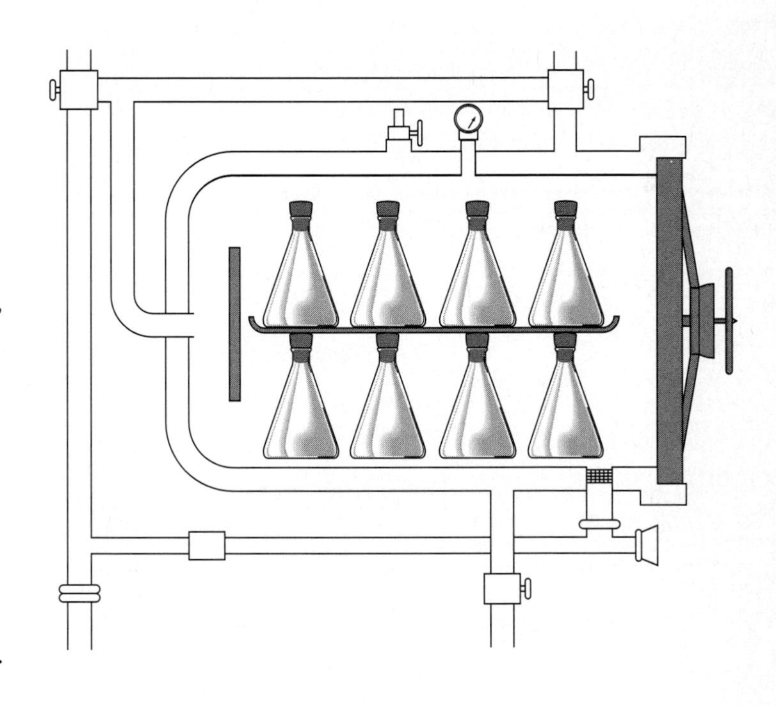

1. Mark the pathway that steam and air take through this autoclave.
2. You can put several objects in an autoclave at one time, but what's wrong with the arrangement shown in the figure?
3. You can wrap small items in paper but not in aluminum foil for autoclaving. Why not?

Figure 9.2

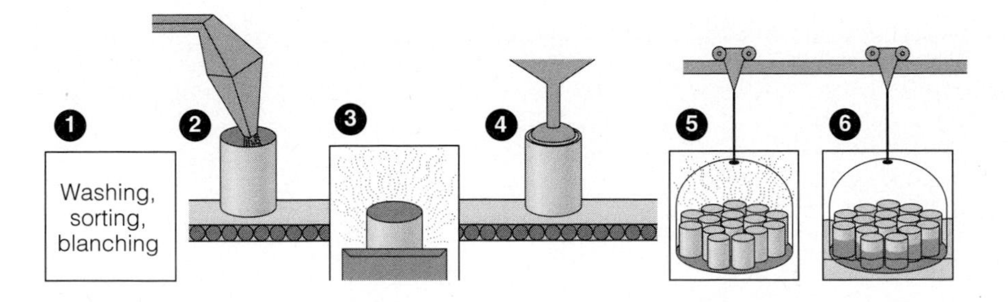

1. Label the steps, and identify the purpose of each step.
2. Why aren't the cans sealed before steaming?

3. Which step is the same as autoclaving?

4. Of all of the bacteria, why are *Clostridium* and *Bacillus* bacteria the most likely to survive this treatment? (*Hint:* Refer to Unit 2.)

5. How can bacteria that survive the canning process grow in a sealed can? (*Hint:* Refer to Unit 3.)

6. If you don't have an autoclave, pressure cooker, or retort, will boiling a solution longer than usual sterilize it?

Figure 9.3

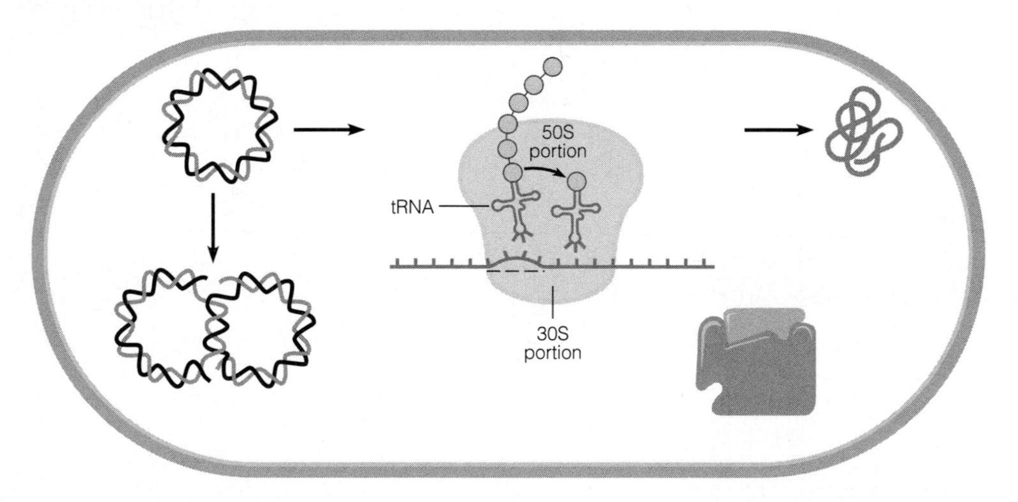

Label the site of action of each of the following antibiotics:

a. Bacitracin
b. Chloramphenicol
c. Erythromycin
d. Penicillins
e. Quinolones
f. Streptomycin
g. Sulfanilamide
h. Tetracyclines
i. Vancomycin

Definitions

Match the following statements to words from Key Terms and Concepts:

1. Equipment used for steam sterilization. ____________________
2. In microbiology, the method of sterilizing an inoculating loop. ____________________
3. 72°C for 15 sec. ____________________
4. 121°C for 15 min. at 15 psi. ____________________
5. 170°C for 2 hr. ____________________
6. Used to sterilize heat-labile solutions such as antibiotics. ____________________
7. Used to sterilize milk. ____________________
8. Use of two drugs interferes with the action of one or both. ____________________

Matching Questions I

Match the following choices to the statements below:

a. Denature protein
b. Oxidizing agent
c. Disrupts plasma membrane
d. No antimicrobial effect

1. Household bleach ____
2. Tincture of iodine ____
3. Mercurochrome ____
4. Hexachlorophene ____
5. Sodium lauryl sulfate (household detergent) ____
6. Cetyl pyridinium chloride ____
7. Alcohol ____

Around Your Home

Read the active ingredients on your household disinfectants and antiseptics, and read about their methods of action in your textbook. Ask students to post what they have found and look for common characteristics such as the most often used compound for bathrooms (against gram-negative bacteria) or in mouth/throat/face products (against gram-positive bacteria).

Are there any preservatives in your foods and cosmetics? How can you tell? By reading labels and by inspection, seeing no microbial growth.

Matching Questions II

Match the following choices to the drugs listed below:

a. Antibacterial
b. Antihelminthic
c. Antiviral

1. Erythromycin ____
2. Niclosamide, prevents ATP generation in mitochondria ____
3. Nucleoside analogs ____
4. Penicillin ____
5. Protease inhibitors ____
6. Streptomycin ____
7. Tetracycline ____

Fill-In Questions

Use the following choices to complete the statements below. Choices may be used once, more than once, or not at all.

260 nm	chlorine	osmotic lysis
500 nm	disinfectants	*Penicillium*
70S	ethylene oxide	phosphatase
80S	fungi	plasmolysis
antibiotics	gram-negative bacteria	sterile
antiseptics	gram-positive bacteria	*Streptomyces*
aseptic	isoniazid	

1. ____________________ techniques are used to reduce contamination.
2. ____________________ are chemicals used to decontaminate nonliving materials.
3. Antibiotics that inhibit bacterial protein synthesis work at ____________________ ribosomes.
4. Penicillin is usually used against ____________________.
5. The presence of ____________________ is used to determine whether milk has been pasteurized.
6. Most antibiotics are made by ____________________.
7. The most effective ultraviolet wavelength for killing bacteria is ____________________.
8. High concentrations of salt and sugar can kill cells by ____________________.
9. ____________________ is a chemical sterilant.
10. ____________________ is used to disinfect drinking water and wastewater.

Short-Answer Questions

1. Which graph more accurately shows the effect of heat treatment on a population of bacteria? Why?

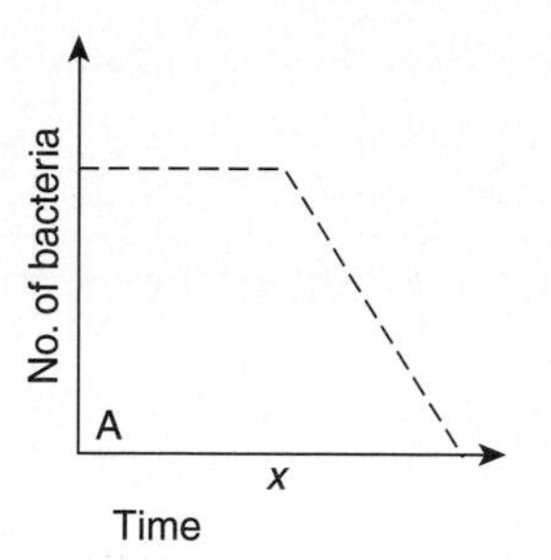

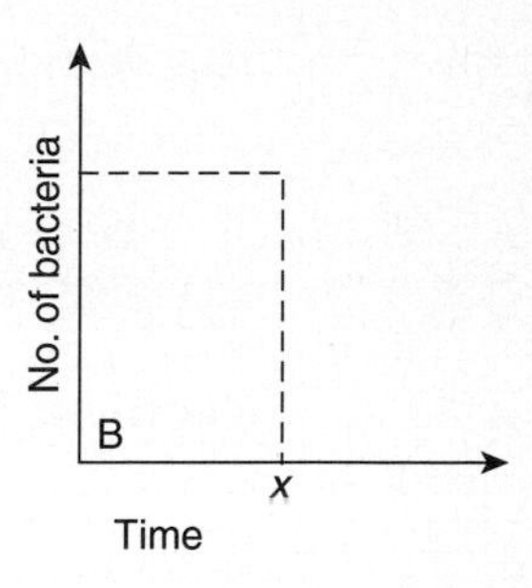

2. The DRT for *S. enteritidis* is 4.5 min. at 58°C and 6.0 min. at 57°C. If you had eggs containing 10^6 *S. enteritidis*, what would happen if you cooked the eggs for 27 min. at 58°C? How long would it take to kill 10^6 bacteria at 58°C? At 57°C? (Read about the importance of *Salmonella enteritidis* contamination of poultry products and a disastrous dinner party on pp. 666–667 in your textbook.)

3. How do the methods of action of antiviral drugs differ from those of antibacterial drugs?

4. How would you determine whether your favorite disinfectant is actually effective?

5. Why do nucleoside analogs stop viral infections? What effect do they have on host cells?

6. Why is 70% alcohol a more effective disinfectant than 100% alcohol?

7. Why was penicillin considered a "magic bullet"?

8. Plants have cell walls; why doesn't penicillin kill plants?

A Bioethics Question

This is an important topic and can provide an opportunity for students to interact with each other while trying to resolve some important issues. You can include a question on the value of canning compared to freezing for developing countries. Students can post their arguments/evidence on an electronic bulletin board to arrive at a class conclusion.

Question: Should chemical preservatives be added to food?

What are the benefits of food preservatives? The risks? Students should indicate that refrigeration is too expensive for developing countries and that loss of food to microbial decomposition is too costly for farmers and consumers in all countries.

Conclusion: What recommendations do you have regarding the use of preservatives?

Study Questions

***Microbiology: An Introduction,* Sixth Edition**

Pages	Review Questions	Multiple Choice Questions	Critical Thinking Questions	Clinical Applications Questions
204–206	All	All	All	All
555–556	All	All	All	All
759	2–4	1, 3–5, 8		2

CD Activities

Do the Chapter 7 and Chapter 20 quizzes.

Web Activities

1. Do the Chapter 7 and Chapter 20 quizzes.
2. Read "Antimicrobial Peptides" and propose a reason that microbial fermentations have been used to preserve food. Bacteriocins and lantibiotics produced by bacteria prevented the growth of pathogenic or spoilage organisms.

ANSWERS

Concept Map 9.1

1. See pp. 184–191 in your textbook.
2. Radiation, Temperature (steam under pressure and hot air), Membrane filtration
3. See pp. 184–191 in your textbook.

4. a. Antiseptic
   b. Antifungal
   c. Antiviral
   d. Antihelminthic
   e. Antibacterial
5. You can't use something that kills all cells, because you would kill the host, too. Therefore, you need to target parts of each of these parasites that are not like the (human) host. Each of these groups differs a little from the other; for example, fungi and bacteria both have cell walls but these cell walls are chemically different, so a drug that affects the synthesis or chemistry of one cell wall will not affect the other.

Figure 9.1

1. See Figure 7.2 (p. 185) in your textbook.
2. The top tray of flasks is placed directly on top of the bottom flasks, which will prevent steam from entering the bottom flasks.
3. Steam will not penetrate most foils, so the wrapped items will only be exposed to heat, not steam.

Figure 9.2

1. See Figure 28.1 (p. 743) in your textbook.
2. The can needs to be open so steam can push air out of the can.
3. Retort, step 5.
4. *Clostridium* and *Bacillus* produce endospores that are heat resistant.
5. The bacteria that grow in sealed cans are growing anaerobically using either fermentation or anaerobic respiration.
6. No, some bacteria are not killed by 100°C.

Figure 9.3

Check your answers with Figures 20.2 (p. 534) and 20.4 (p. 535) in your textbook.

Definitions

1. autoclave
2. flaming
3. high-temperature short-time pasteurization
4. autoclave
5. hot-air sterilization
6. membrane filters
7. ultra-high-temperature
8. antagonism

Matching Questions I

1. b
2. a
3. a
4. a, c
5. d
6. c
7. a, c

Matching Questions II

1. a
2. b
3. c
4. a
5. c
6. a
7. a

Fill-In Questions

1. aseptic
2. disinfectants
3. 70S
4. gram-positive bacteria
5. phosphatase
6. *Streptomyces*
7. 260 nm
8. plasmolysis
9. ethylene oxide
10. chlorine

Short-Answer Questions

1. A; the cells do not all die at the same time; death is logarithmic.
2. One cell remains after 27 min. at 58°C. Cooking longer than 27 min. at 58°C and longer than 36 min. at 57°C.
3. Antibacterial drugs target prokaryotic structures such as the peptidoglycan cell wall or 70S ribosomes. Antiviral drugs are usually mutagens. New drugs are targeting virus-specific enzymes.
4. The disk-diffusion test would be a quick screen. The use-dilution test is a more realistic test.
5. Analogs stop DNA synthesis or cause mutations. If host cell enzymes use the analogs, they would have the same effects on growing cells.
6. Denaturation of (water-soluble) proteins requires water.
7. Penicillin killed bacteria but had no effect on human cells.
8. Plant cell walls are made of cellulose; penicillin interferes with peptidoglycan synthesis.

Unit 10

MICROBIAL INTERACTIONS

There are more animals living in the scum on the teeth in a man's mouth than there are men in a whole kingdom.

—ANTONI VAN LEEUWENHOEK, 1683

LEARNING OBJECTIVES

★ 1. List the defining characteristics of fungi. p. 321

2. Differentiate between asexual and sexual reproduction, and describe each of these processes in fungi. p. 321

3. List the outstanding characteristics of the six divisions of algae discussed in Chapter 12. p. 322

4. List the defining characteristics of the four divisions of fungi described in Chapter 12. p. 325

5. List the distinguishing characteristics of lichens, and describe their nutritional needs. p. 325

6. Describe the roles of the fungus and the alga in a lichen. p. 325

7. Compare and contrast cellular slime molds and plasmodial slime molds. p. 327

8. List the defining characteristics of algae. p. 331

9. Identify the environmental needs for algae. p. 332

10. Define normal and transient microbiota. p. 395

★ 11. Compare commensalism, mutualism, and parasitism, and give an example of each. p. 395

12. Compare and contrast normal and transient microbiota with opportunistic microorganisms. p. 396

★ 13. Define mycorrhiza. p. 715

★ 14. Define symbiosis, differentiate between parasitism and mutualism, and give an example of each. p. 715

★ 15. Define bioremediation. (Also in Unit 8.) p. 724

★ 16. Give two examples of the use of bacteria to remove pollutants. (Also in Unit 8. You might ask for two examples not listed in Unit 8.) p. 724

READING

Chapter 12 (pp. 320–338), Chapter 14 (pp. 395–398), Chapter 25 (pp. 658–659), Chapter 27 (pp. 715, 723–724, 726).

SUGGESTED LABS FROM JOHNSON AND CASE, *LABORATORY EXPERIMENTS IN MICROBIOLOGY*, FIFTH EDITION

Exercise 33: Fungi: Yeasts
Exercise 34: Fungi: Molds
Exercise 35: Phototrophs: Algae and Cyanobacteria
Exercise 56: Microbes in Soil: Bioremediation

KEY TERMS AND CONCEPTS

algal blooms p. 325
algin p. 333
arbuscules p. 715
arthrospore p. 324
ascospore p. 326
asexual spores p. 323
bacteriocins p. 396
basidiospores p. 326
bioremediation p. 724
blades p. 331
budding yeast p. 322
chlamydospore p. 324
coenocytic hyphae p. 321
cofactors p. 398
commensalism p. 396
conidiospore p. 324
cortex p. 336
cytoplasmic streaming p. 338
dimorphism p. 323
diploid zygote p. 325
eyespot p. 334
fission yeasts p. 322
haploid p. 325
holdfasts p. 331
hyphae p. 321
lichen p. 335
medulla p. 336
microbial antagonism p. 395
mitosis p. 324
mutualism pp. 396, 715
mycelium p. 322
mycology p. 320
mycorrhizae pp. 320, 715
normal flora p. 395
normal microbiota p. 395
parasitism pp. 396, 715
pellicle p. 334
phytoplankton p. 726
plankton p. 333
plasmodium p. 337
pseudohypha p. 322
rhizines p. 336
septa p. 321
septate hyphae p. 321
sexual spores p. 324
slime mold p. 337
sporangiospore p. 325
sporangium p. 325
spores p. 323
stipes p. 331
symbiosis pp. 396, 715
synergism p. 398
thallus p. 321
transient microbiota p. 395
vegetative p. 321
vesicles p. 715
zygospore p. 326

INTRODUCTION

Be sure to read the Study Outline for Chapter 12.

As you learned in Unit 8, microorganisms live in symbiotic associations with all plants and animals. Some microorganisms can cause disease of these organisms, and others live in commensal or mutualistic relationships. Lichens (pp. 335–336) are an example of a symbiotic relationship between an alga and a fungus. Some microbes such as mycorrhizae (p. 715) and nitrogen-fixing bacteria live in association with plant roots. Animals cannot digest the carbohydrate cellulose. However, herbivores such as cows and termites eat only plants, and cellulose is the bulk of the carbohydrates they consume. These animals house symbiotic bacteria in their digestive systems. The bacteria produce the enzyme cellulase to digest the cellulose for their growth; the host animal also gets metabolites from the bacterial enzymes. Refer to Unit 3.

Microbes are essential for our health. Some microbes, especially bacteria, that live on and in the human body provide essential growth factors; and other microbes produce bacteriocins, which prevent the growth of pathogens. Other microbes protect us from disease by using nutrients, which prevents pathogens from growing. These microbes that benefit us while we provide nutrients are in a mutual relationship with us, their hosts. There are also commensal microbes that live harmlessly on us; they benefit by having a warm place to live, and we are not affected.

In Unit 1, you learned about a food chain. A food chain is drawn as a straight line. Actually, the relationships are more interrelated, so the lines (representing the flow of energy and nutrients) crisscross to form a web. Few ecosystems are so simple that they are characterized by a single, unbranched food chain. Several types of primary consumers usually feed on the same plant, and one species of primary consumer may eat several different plants. As you saw in Unit 1, decomposers feed on the remains of organisms in every level in the food chain and recycle the elements back to forms that plants can use. And some microbes, like the nitrogen-fixing bacteria, occupy specialized niches in the nutrient cycles.

Some microorganisms are pathogenic—that is, they cause disease when they are in an appropriate host. Human diseases and defenses against diseases are the subjects of Units 11 and 12.

VIDEO: MICROBIAL INTERACTIONS

Terms

The following new terms are introduced in this video:

anoxic
dichloroethylene
fistula
mycologist
pregastric
trichloroethylene
web of life

Preview of Video Program

This video program begins with an introduction to the commensal and mutual microorganisms living with animals.

A human body provides a suitable environment for bacteria; consequently, a microbial ecosystem exists in and on the human body. This ecosystem functions like the environmental ecosystems studied in Unit 8. In the human large intestine, bacteria use food that its human host did not use and provide vitamins that the human can use.

At Michigan State University, John Breznak demonstrates the use of a fistula or door into the cow's rumen to study microbes. The rumen is at the base of the esophagus, just before the stomach (pregastric). These microbes must grow anaerobically because the rumen is an anoxic (without oxygen) environment. As you saw in Unit 8, the cow takes in cellulose in the leaves it eats; however, the cow cannot digest cellulose. The cow relies on bacteria in a specialized structure called a rumen. These bacteria have the enzyme cellulase, which can break down cellulose.

Breznak also shows some protozoa in the gut of a termite. Termites, like cows, ingest food that they cannot digest. In Unit 7, you saw the bacteria that are also in the termite gut. These microbes digest the cellulose and wood for the termite.

In the Costa Rican rain forest, how many species of microbes are in a teaspoon of soil? 5000 Leaf-cutter ants are seen cutting pieces of leaves to bring to their nests. In the nests, fungi grow on the leaves and decompose them. Like cows and termites, ants cannot digest cellulose. Unlike cows and termites, these plant-eating ants do not house cellulose-degrading microbes in a specialized digestive organ. What do leaf-cutting ants use for food? fungi You will see time-lapse photography of a slime mold growing (~12 min., 15 sec.).

Ignacio Chapela, a mycologist, studies fungi in the rain forest. The primary role of these fungi is to decompose the cellulose and wood, to return CO_2 to the atmosphere to be used by plants and other producers. Some fungi are more intimately associated with plants—like the cow and its bacteria. Fungi that grow symbiotically in or on the roots of a plant are called mycorrhizae. A plant normally produces sugars for itself, but in a symbiotic relationship with fungi, the plant secretes some sugars that the fungus can use. And the fungus absorbs soil nutrients that the plant can use.

As you saw in Unit 8, the use of microbes to degrade or detoxify pollutants is called bioremediation. In this video, Terry Hazen uses naturally occurring bacteria to clean up chlorinated organic solvents from contaminated soil at the Savannah River nuclear weapons site.

Video Questions

1. Why can microbes be described as "masters of life on Earth"? Microbial activities directly provide nutrients such as O_2, organic acids, and vitamins for other organisms. And, through decomposition, microbes recycle organic matter to make elements available for plant growth.

2. Does the cow actually do the pregastric fermentation? No.

3. Methane is produced in the cow's rumen. What organisms use the methane? Methanotrophic bacteria.

4. What did Terry Hazen pump into the ground at the Savannah River site? Why did he add methane if he wanted the bacteria to use the chloroethylene compounds? Air and methane to enrich for methanotrophs that use the methane. The methane was carefully measured so that the bacteria would increase in number then need more carbon and switch to using the pollutants.

5. What percentage of O_2 in the atmosphere is produced by terrestrial plants? 50% By microorganisms? 50% Which microbes produce O_2? Cyanobacteria and phytoplankton.

Concept Map 10.1

This map shows terms related to the web of life. The arrows represent the flow of energy and nutrients.

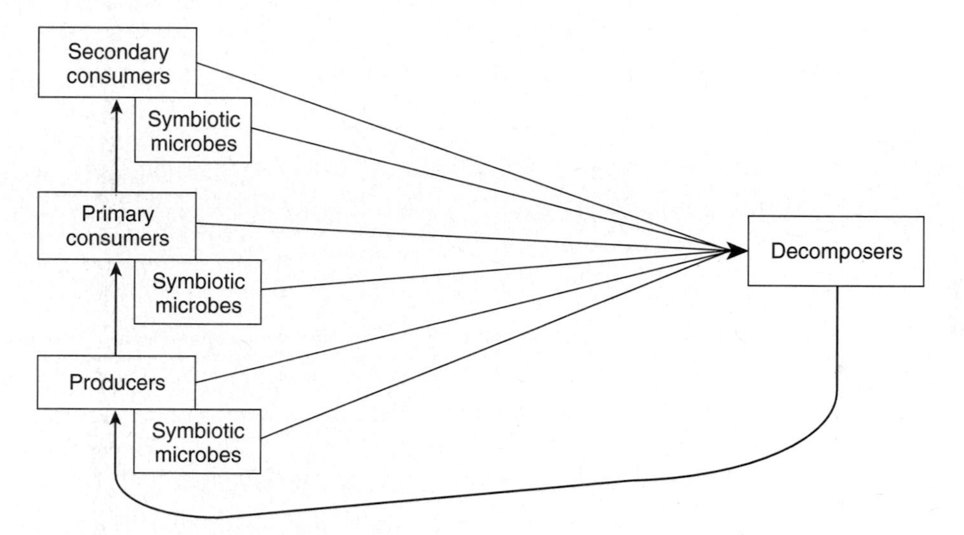

1. Add arrows between producers and their symbiotic microbes.
2. Where would you put parasites in this food web?

Figure 10.1

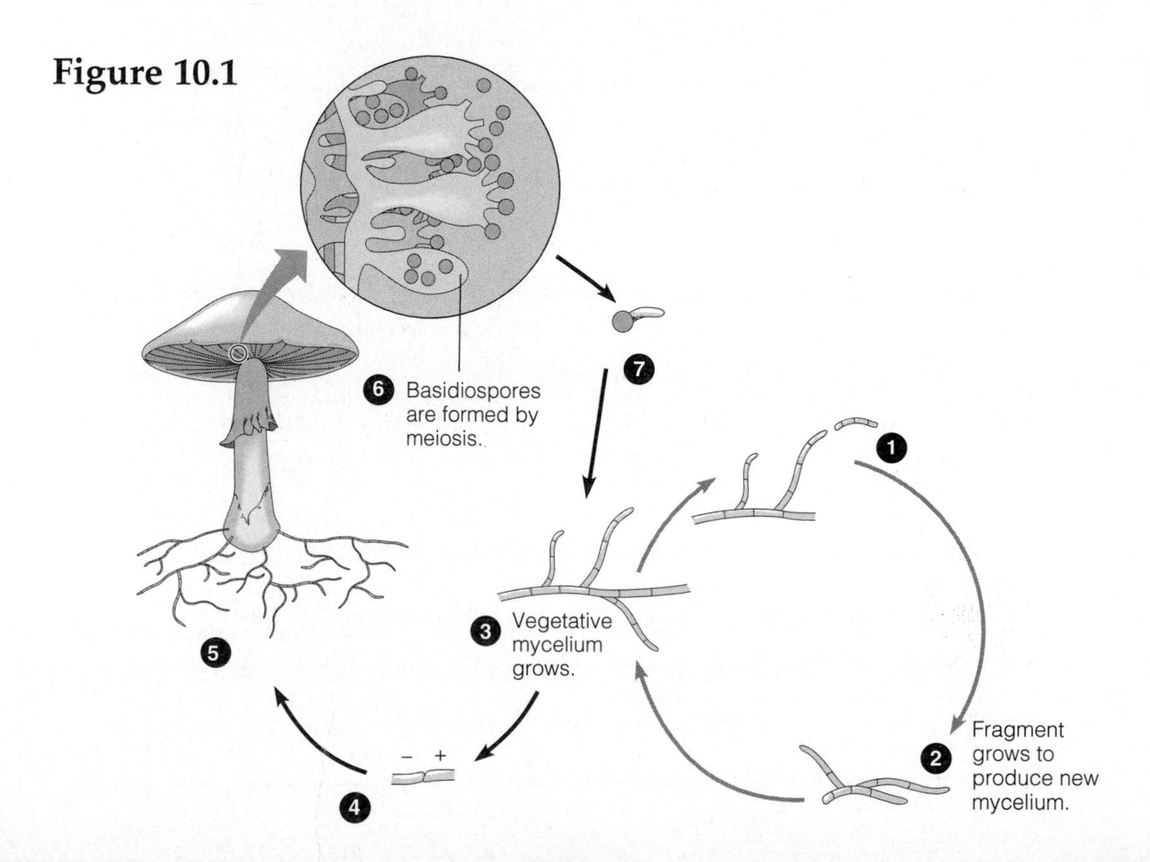

1. Label the hyphae, gills, and spores.
2. Label a germinating spore.
3. Identify the sexual and asexual phases of growth.

Figure 10.2

Place the following terms in the cow:

cellulase
cellulose
CO_2
fatty acids
fermentation
glucose
methane

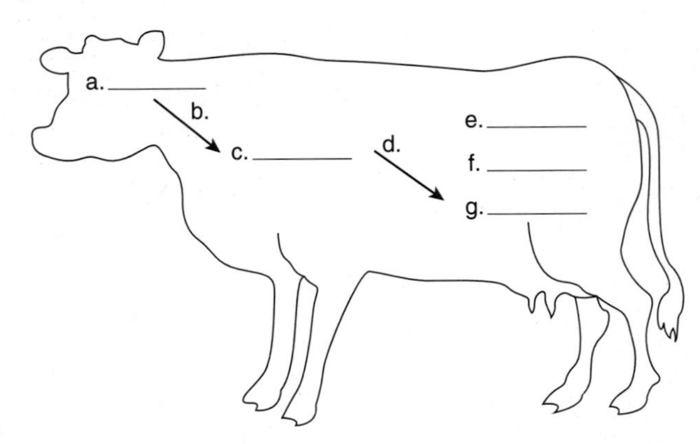

1. In the video, Jim Leinfelder refers to a catalyst to break down cellulose. What is the catalyst? ______________________
2. Identify the microbial processes. ______________________
3. Is this food chain aerobic or anaerobic? ______________________
4. Why do cows chew their food more than cats and dogs do?

5. How does the cow get nitrogen (as —NH_2) in order to make proteins, such as muscles and enzymes?

Definitions

Match the following statements to words from Key Terms and Concepts:

1. Microbes associated with plant roots that help plants absorb nutrients. ______________
2. The type of sexual spores produced by most mushrooms. ______________________
3. The type of asexual spore produced by *Penicillium.* ______________________
4. Sexual spore produced by molds in the Ascomycota. ______________________

5. Fungal structures inside plant roots that provide nutrients to the plant.
   ____

6. The microbes living more or less permanently in and on the human body.
   ____

7. Free-floating organisms. ______________________

8. Intracellular movement of protoplasm. ______________________

9. Structures used by sessile (nonplanktonic) algae to attach to a rock. ______________________

Comparison Questions

Compare and contrast each of the following pairs of terms:

1. Slime mold and fungus

2. Fungus and bacterium

3. Cellular slime mold and plasmodial slime mold

4. Mold and mushroom

5. Mutualism and commensalism

6. Lichen and fungus

Make a Food Web

In Unit 8, you saw a salt-marsh food web that consisted of the following organisms:

Bacteria
Crab larva
Fish
Great Blue Heron
Spartina

Draw a food web showing the interrelationships between and among these organisms.

Symbiosis Questions I

Match the choices below to the following pairs:

commensalism
mutualism
parasitism
none of the above

1. Termite and the microbes in its gut ______________________
2. Peas and root nodule bacteria ______________________
3. *Agrobacterium* and cassava (Unit 5) ______________________
4. *E. coli* and humans ______________________
5. Plant and arbuscules ______________________
6. *Spartina* and bacteria on its surface (Unit 8) ______________________
7. Methanotrophs and trichloroethylene ______________________
8. Leaf-cutting ants and cellulose-degrading fungi ______________________
9. Heron and fish (Unit 8) ______________________
10. Alga and fungus in a lichen ______________________

Symbiosis Questions II

In the following pairs, which one of each pair cannot live without the other one?

1. Termite and the microbes in its gut ______________________
2. Peas and root nodule bacteria ______________________
3. *Agrobacterium* and cassava (Unit 5) ______________________
4. *E. coli* and humans ______________________
5. Plant and arbuscules ______________________

6. *Spartina* and bacteria on its surface (Unit 8) ____________________
7. Methanotrophs and trichloroethylene ____________________
8. Leaf-cutting ants and cellulose-degrading fungi ____________________
9. Heron and fish (Unit 8) ____________________
10. Alga and fungus in a lichen ____________________

Short-Answer Questions

1. Do bacteria only grow anaerobically in an anoxic environment? (*Hint*: Review Unit 3.)

2. Why can strict aerobic bacteria grow next to strict anaerobic bacteria in soil?

3. Was the slime mold shown in the rain forest a cellular or plasmodial slime mold (~12 min., 15 sec.)? How can you tell?

4. Are bacteria producers or decomposers? Briefly explain.

5. Why are the normal microbiota on the human body sometimes called normal *flora*?

6. Chapelo said fungi are more closely related to animals than bacteria. What information would lead to this conclusion? (*Hint:* Refer to Unit 6.)

7. Why are fungi classified as microbes while most other multicellular organisms are not?

Hypothesis Testing

This question can be set up on an electronic bulletin board so that each student who responds has to add the next step or amend the previous step in the procedure.

Hypothesis: You suspect the new pine tree (*Pinus radiata*) you planted in your yard is not doing well because its mycorrhizae are missing. Design an experiment to test your hypothesis.

Your Experimental Design: Place one plant in sterile soil, a second plant in your test soil, and a third in soil from the tree's natural habitat. Assuming the plant grows best in its native soil, look for fungi in the soil and associated with the roots of the plant. You could try looking for hyphae associated with the roots of healthy plants, and absent from chlorotic, poorly growing plants, but the presence of fungi doesn't confirm their necessity.

Conclusion: What results would confirm your hypothesis? Briefly explain.

Study Questions

Microbiology: An Introduction, Sixth Edition				
Pages	Review Questions	Multiple Choice Questions	Critical Thinking Questions	Clinical Applications Questions
356–358	1–9		2	2, 4
740	2			

CD Activity

Do the Chapter 12 quiz.

Web Activity

Do the Chapter 12 quiz.

Outdoor Activity

1. Find a mushroom and carefully remove the soil or plant matter at its base. What are the white hairlike structures called? hyphae ____ How far do they extend from the mushroom? ____________ What is the function of these structures? absorb and metabolize food ____ What is the purpose of the mushroom? reproduction, produces spores ____________

2. Does your mushroom have gills? ____________ Pores? ____________ If so, diagram the structure.

Mushrooms are identified by macroscopic appearance and microscopic appearance of the spores. Remove the stipe and place the mushroom cap on a piece of paper for a few hours. Remove the cap from the paper. If the mushroom had colored spores, you will see a "spore print" on white paper. Sketch the spore print. How closely does it resemble your diagram of the gills or pores? Why? The spore print should look like the gill/pore picture because the basidiospores were produced on the inside surfaces of the gills/pores.

ANSWERS

Concept Map 10.1

1. Producers $\rightleftharpoons$ Symbiotic microbes.
2. Parasites will obtain food and energy from any of the organisms in the web. For example, a deer (primary consumer) may have a parasitic tick sucking the deer's blood. The tick may have a parasitic mite, which could have a parasite fungus; and the fungus could be infected by a virus.

Figure 10.1

Check your answers against Figure 12.8 (p. 330) in your textbook.

Figure 10.2

a. cellulose
b. cellulase
c. glucose
d. fermentation
e. fatty acids
f. methane
g. CO_2

1. The enzyme cellulase.
2. All of those shown.
3. Anaerobic.
4. Cows grind up the cellulose fibers, which makes it easier for bacteria to hydrolyze the cellulose.
5. Nitrogen-fixing bacteria in the rumen provide nitrogen for the bacteria to assimilate into protein. Some bacteria will be moved to the small intestine, where they can be digested by the cow.

Definitions

1. mycorrhizae
2. basidiospores
3. conidiospore
4. ascospore
5. arbuscules
6. normal microbiota
7. plankton
8. cytoplasmic streaming
9. holdfasts

Comparison Questions

1. Both are eukaryotic, chemoorganotrophs. Fungi may be unicellular or multicellular; multicellular fungi form mycelia. Slime molds may be (1) unicellular amoebae that aggregate to form asexual spores or (2) giant multinucleated cells.
2. Members of both groups are important decomposers (chemoorganotrophs). Fungi are eukaryotic and all are chemoorganotrophs. Bacteria are prokaryotic and may be organotrophs or lithotrophs.
3. Both are eukaryotic. The vegetative form of a cellular slime mold is a uninucleated amoeba. The vegetative form of a plasmodial slime mold is a multinucleated cell.
4. Both are multicellular eukaryotes. Molds consist of a mycelium. In mushrooms, some of the mycelia are matted together to form the macroscopic mushroom that produces sexual spores.
5. Both are symbiotic relationships. In mutualism, the two organisms benefit from each other. In commensalism, one organism benefits and the other is unaffected.
6. Lichens consist of algae and fungi in a symbiotic relationship. Lichens are photoautotrophs. Fungi are eukaryotic organotrophs.

Make a Food Web

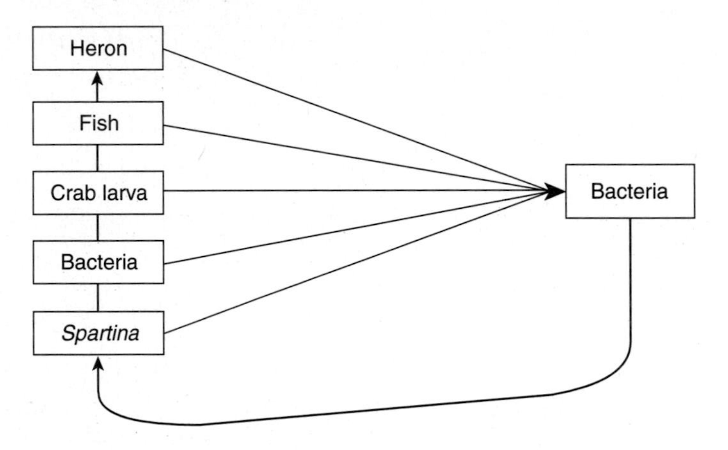

Symbiosis Questions I

1. mutualism
2. mutualism
3. parasitism
4. mutualism
5. commensalism
6. commensalism
7. none of the above
8. mutualism
9. none of the above
10. mutualism

Symbiosis Questions II

1. Termite
2. Peas
3. Neither
4. Humans
5. Plant (some of the fungi)
6. Both
7. The microbe is not dependent on trichloroethylene
8. Ants
9. Heron (although it may have alternative foods such as frogs and mice)
10. If the two are separated, the lichen no longer exists. Both reproduce separately and can be grown in vitro.

Short-Answer Questions

1. No, some bacteria grow anaerobically even when they are in the presence of O_2. For example, *Streptococcus* lacks an electron transport chain and can only grow by fermentation.
2. Aerobes might grow and use up the oxygen, thus allowing the anaerobic bacteria to grow.
3. Plasmodial slime mold. The growing plasmodium was shown. Cellular slime molds are unicellular except when they are forming the slug and stalk.
4. Both. Some bacteria are autotrophs that produce O_2 and fix CO_2; others are organotrophs that decompose organic matter.
5. Recall from Unit 6 that bacteria were once thought to be plants.
6. Fungi are eukaryotic, which puts them in the Eukarya. Recent rRNA analysis shows that fungi are more closely related to animals than to the other Eukarya.
7. The cells in a fungus do not form tissues with specialized functions like the blood of an animal or vascular tissue of a plant.

Unit **11**

HUMAN DEFENSES

Smallpox, so fatal and so general amongst [the English], is here [in Turkey] entirely harmless by the invention of ingrafting.

—MARY MONTAGU, 1717

LEARNING OBJECTIVES

★	1.	Describe the structure of the skin and mucous membranes and the ways pathogens can invade the skin.	p. 558
	2.	Provide examples of normal skin microbiota, and state the general locations and ecological roles of its members.	p. 559
★	3.	Identify the principal portals of entry.	p. 421
	4.	Define LD_{50} and ID_{50}.	p. 423
	5.	Using examples, explain how microbes adhere to cells.	p. 423
★	6.	Explain how capsules and cell wall components contribute to pathogenicity.	p. 425
	7.	Describe the effects of leukocidins, hemolysins, coagulases, kinases, hyaluronidases, and collagenase.	p. 426
	8.	Contrast the nature and effects of exotoxins and endotoxins.	p. 427
★	9.	Describe the role of the skin and mucous membranes in nonspecific resistance.	p. 440
★	10.	Differentiate between mechanical and chemical factors, and list five examples of each.	p. 440
★	11.	Define phagocytosis and phagocyte.	p. 443
★	12.	Classify phagocytic cells, and describe the roles of granulocytes and monocytes.	p. 443
	13.	Define differential white blood cell count.	p. 443
★	14.	Describe the process of phagocytosis, and include the stages of adherence and ingestion.	p. 425
★	15.	List the stages of inflammation.	p. 449
★	16.	Describe the roles of vasodilation, kinins, prostaglandins, and leukotrienes in inflammation.	p. 449
★	17.	Describe phagocyte migration.	p. 449
	18.	Define the cause and effects of fever.	p. 451
	19.	List the components of the complement system.	p. 453
	20.	Describe two pathways of activating complement.	p. 453
	21.	Describe three consequences of complement activation.	p. 453
	22.	Define interferons.	p. 456
	23.	Compare and contrast the actions of α-IFN and β-IFN with γ-IFN.	p. 456
★	24.	Differentiate between immunity and nonspecific immunity.	p. 461
★	25.	Contrast the four types of acquired immunity.	p. 461
	26.	Differentiate between humoral (antibody-mediated) and cell-mediated immunity.	p. 463

★ 27. Define antigen and hapten. p. 463

28. Explain the function of antibodies, and describe their structural and chemical characteristics. p. 464

29. Name one function for each of the five classes of antibodies. p. 464

30. Name the function of B cells. p. 467

31. Define apoptosis, and give a potential medical application. p. 467

32. Describe the clonal selection theory. p. 468

33. Explain how an antibody reacts with an antigen, and identify the consequences of the reaction. p. 469

★ 34. Distinguish between a primary and a secondary immune response. p. 471

35. Define monoclonal antibodies, and identify their advantage over conventional antibody production. p. 471

36. Identify at least one function of each of the following in cell-mediated immunity: cytokine, interleukin, interferons. p. 472

★ 37. Describe at least one function for each of the following: T_H cell, T_C cell, T_D cell, T_S cell, APC, MHC, activated macrophage, NK cell. p. 474

★ 38. Compare and contrast cell-mediated and humoral immunity. p. 475

★ 39. Compare and contrast T-dependent antigens and T-independent antigens. p. 478

40. Describe the role of antibodies and NK cells in antibody-dependent cell-mediated cytotoxicity. p. 478

★ 41. Define vaccine. p. 485

★ 42. Explain why vaccination works. p. 486

43. Define herd immunity. p. 486

★ 44. Differentiate between the following, and provide an example of each: attenuated, inactivated, toxoid, subunit, and conjugated vaccines. p. 488

45. Contrast subunit vaccines and nucleic-acid vaccines. p. 488

46. Compare and contrast the production of whole-agent vaccines, recombinant vaccines, and DNA vaccines. p. 490

READING

Chapter 15 (pp. 420–430), Chapter 16, Chapter 17, Chapter 18 (pp. 485–490), and Chapter 21 (pp. 558–560).

SUGGESTED LABS FROM JOHNSON AND CASE, *LABORATORY EXPERIMENTS IN MICROBIOLOGY*, FIFTH EDITION

Exercises 41–44: Immunology

KEY TERMS AND CONCEPTS

Microbial Mechanisms of Pathogenicity

adherence pp. 423, 446
adhesins p. 423
endotoxin p. 429
exotoxin p. 428
invasins p. 427
ligands p. 423
parental route p. 423
receptors p. 423
toxins p. 428

Nonspecific Defenses of the Host

abscess p. 449
antiviral proteins (AVPs) p. 457
basophils p. 444
ciliary escalator p. 442
complement p. 453
complement fixation p. 456
crisis p. 452
cytokines pp. 445, 472
cytolysis p. 453
dermis p. 440
differential white blood cell count p. 443
emigration p. 449
eosinophils p. 444
epidermis p. 440
fever p. 449
fixed macrophages p. 445
formed elements p. 443
granulocytes p. 443
histamine p. 449
immunity pp. 440, 461
inflammation pp. 449, 471
interferons (IFNs) pp. 456, 472
keratin p. 440
kinins p. 449
lacrimal apparatus p. 441
leukocytes p. 443
leukotrienes p. 449
lymphocytes p. 445
lysozyme p. 442
macrophages p. 444
margination p. 449
membrane attack complex p. 453
monocytes p. 444
mononuclear phagocytic system p. 445
neutrophils p. 443
nonspecific resistance p. 440
opsonization pp. 446, 470
phagocytes p. 443
phagolysosome p. 447
phagosome p. 447
plasma p. 443
prostaglandins p. 449
pus p. 449
sebum p. 442
serum pp. 453, 462
shivering p. 452
specific resistance p. 440
transferrins p. 442
transmembrane channels p. 453
vasodilation p. 449
wandering macrophages p. 445

B Cells and Humoral Immunity

acquired immunity p. 461
agglutination p. 469
anamnestic response p. 471
antibodies p. 464
antibody-dependent cell-mediated cytotoxicity (ADCC) pp. 470, 478
antibody-mediated immunity p. 463
antibody titer p. 472
antigen-antibody complex p. 469
antigen-binding sites p. 464
antigenic determinants p. 464
antigen-presenting cells (APCs) p. 474
antigen receptors p. 467
antigens p. 463
antiserum p. 462
apoptosis p. 467
artificially acquired active immunity p. 462
artificially acquired passive immunity p. 462
B cells p. 463
CD4 cells p. 474
CD8 cells p. 474
cell-mediated immunity p. 463
chemokines p. 473
chimeric monoclonal antibodies p. 472
clonal deletion p. 468
clonal selection p. 468
colony-stimulating factor (CSF) p. 472
cytotoxic T (T_C) cells p. 476
delayed hypersensitivity T (T_D) cells p. 476
dendritic cells p. 475
epitopes p. 464
gamma globulin p. 462
globulins p. 462
haptens p. 464
helper T (T_H) cells p. 475
humoral immunity p. 463
hybridoma p. 472
IgA p. 467
IgD p. 467
IgE p. 467
IgG p. 465
IgM p. 466
immune serum globulin p. 462
immunization p. 462
immunoglobulins (Igs) p. 464
immunotoxin p. 472
innate resistance p. 461
interleukins p. 472
lysis p. 472
major histocompatibility complex (MHC) p. 475
memory cells p. 471
memory response p. 471
monoclonal antibodies p. 472
monomer p. 464
natural killer (NK) cells p. 477
naturally acquired active immunity p. 462
naturally acquired passive immunity p. 462
neutralization p. 470
plasma cells p. 467
primary response p. 471
secondary response p. 471
self-tolerance p. 468
serology p. 462
stem cells p. 467
suppressor T (T_S) cells p. 476
T cells p. 463
T-dependent antigen p. 478
T-independent antigens p. 478
tumor necrosis factor (TNF) p. 472
vaccination p. 462
vaccine pp. 462, 485
valence p. 464

Vaccines

acellular vaccines p. 489
attenuated whole-agent vaccines p. 488
conjugated vaccines p. 489
herd immunity p. 486
inactivated whole-agent vaccines p. 488
nucleic-acid vaccines p. 489
recombinant vaccines p. 489
subunit vaccines p. 489
toxoids p. 488
variolation p. 485

INTRODUCTION

Be sure to read the Study Outlines for Chapter 16, pp. 458–459, Chapter 17, pp. 480–483, and Chapter 18, p. 501.

You learned in Unit 10 that there is a microbial ecosystem living in and on humans. The species and numbers of microorganisms that grow on humans are controlled by the microorganisms and environmental conditions. Some microorganisms make antimicrobial compounds such as bacteriocins that prevent other microorganisms from growing. And the skin secretes salt and sebum, which prevent microbial growth. Body tissues are sterile. However, there are microbial populations in several organ systems that have access to the external environment. There are normal microbial populations in the mouth, upper respiratory tract, and the intestines. Microorganisms are kept from entering body tissues because of overlapping layers of cells in the epidermis. There are openings in the skin such as sweat ducts, eyes, ears, nose, and urethra. Generally, microorganisms cannot enter the body through these openings because the flow of fluids is outward. Sometimes a microorganism does enter the body. If host cells are damaged through some trauma, the damaged cells release prostaglandins that attract white blood cells called phagocytes. The phagocytes should engulf and digest the invading microorganism.

Microorganisms that survive this far will elicit an immune response from the host. The immune response is specifically targeted against the particular species and strain of microorganism that is now colonizing in the host. The immune response involves the production of antibodies by plasma cells. Antibodies are proteins made in response to an antigen, that can react with that antigen. An antigen is a molecule that elicits an antibody response. The antigen in this case is one or more large molecules of the microorganism. Antibodies combine with the antigen and can cause (1) phagocytosis of the antigen, (2) lysis of the antigen, or (3) inactivation of the antigen. Antibodies combine with soluble antigens such as viruses or toxins and prevent their attachment to receptors, thereby inactivating the virus or toxin.

B cells that differentiate into plasma cells can produce antibodies against T-independent antigens. B cells need stimulation from T_H cells to respond to some antigens, called T-dependent antigens. The cellular immune system attacks intracellular pathogens and cancer cells.

Vaccines work like an invading antigen to cause an antibody response but do not cause the disease. Vaccines may be dead microbes, attenuated microbes, or subunits of microbes. The earliest recorded use of a vaccine was reported by Mary Montagu in her letters from Turkey. As described in the opening quote for this unit, the process of scraping material from smallpox lesions into a healthy person was called ingrafting, or variolation. The result, if the person didn't develop smallpox, was a mild infection that resulted in antibody production and lifelong immunity to smallpox.

Some microbes have evolved adaptations that allow them to survive in our bodies. These microbes avoid phagocytosis, avoid antibodies, or grow inside a host cell. The role of microbes in human diseases is the subject of Unit 12.

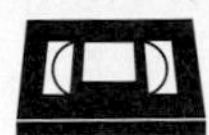

VIDEO: HUMAN DEFENSES

Study Concept Map 11.1 before watching this video, to become familiar with the various cells of the immune system.

Preview of Video Program

In this video program, you will see microorganisms that are normal or transient microbiota on the human skin. The skin is in constant contact with the environment and microorganisms, so it makes sense that some organisms would have evolved to occupy this available niche. Most of the normal and transient microbes are beneficial, but a human may encounter pathogens or opportunistic pathogens. Opportunistic pathogens do not normally cause disease but might if introduced into a new location, such as through a cut. Microbial growth or the products of microbial growth cause disease in a host.

The video and your textbook use a battlefield analogy to describe the relationship between host and microorganism. The skin and mucous membranes are called the First Line of Defense against invading microorganisms. Microorganisms that get through a break in the skin face the Second Line of Defense, phagocytes. Macrophages and neutrophils are phagocytes. At ~11 min., you will see time-lapse photography of a phagocyte ingesting *Candida*.

Sketch the phagocytes in the space provided.

What is *Candida*? The opportunistic yeast that causes infections.

How does the phagocyte move? Pseudopods, amoeboid motion.

Label the *Candida*.
~11 min., 30 sec.

Macrophage ingests *Candida*.

Phagocytes usually digest microbes. Microbes that avoid phagocytosis cause the body to mount its Third Line of Defense, the immune response. At ~13 min., a macrophage does not ingest a pair of cocci because the two cocci are surrounded by a capsule.

Sketch the macrophage in the space provided.

Label the bacteria and the capsule.
~13 min., 30 sec.

Phagocytic cell unable to defeat encapsulated bacterium.

The immune response is usually initiated by macrophages. After phagocytizing and digesting a microbe, macrophages can activate a type of lymphocyte called a helper T cell. The macrophage with a microbial antigen on its surface is called an antigen-presenting cell, which activates a helper T cell. The helper T cell will, in turn, activate B cells to differentiate into antibody-producing plasma cells.

The first and second defenses are considered nonspecific because they are not responding to any particular microorganism. Immunity is a specific defense, against the microorganism that has "made it through the lines."

The immune response means that antibodies will be produced against the microbial antigen or T cells will be activated to react to abnormal cells. A B cell capable of producing the desired antibody is selected and begins to divide. Some of the progeny will be antibody-producing plasma cells. Other progeny will form memory cells that can respond quickly when that antigen is encountered again.

If a microbe is going to successfully live in a human host, it needs to avoid the body's defenses. Stanley Falkow's video micrograph shows *Salmonella* entering host cells, consequently avoiding phagocytosis. You also see this in Figure 15.2 (p. 427) in your textbook. *Salmonella* has evolved a mechanism to live inside a host without killing the host. Read about the different strategies used by successful pathogens in the box on p. 422 in your textbook.

Other lymphocytes are also involved in the immune response. You will see micrographs of cytotoxic T cells (or lymphocytes) in the video. Sketch the cells below. Using your textbook, differentiate between cytotoxic T lymphocytes and killer (non T) lymphocytes.

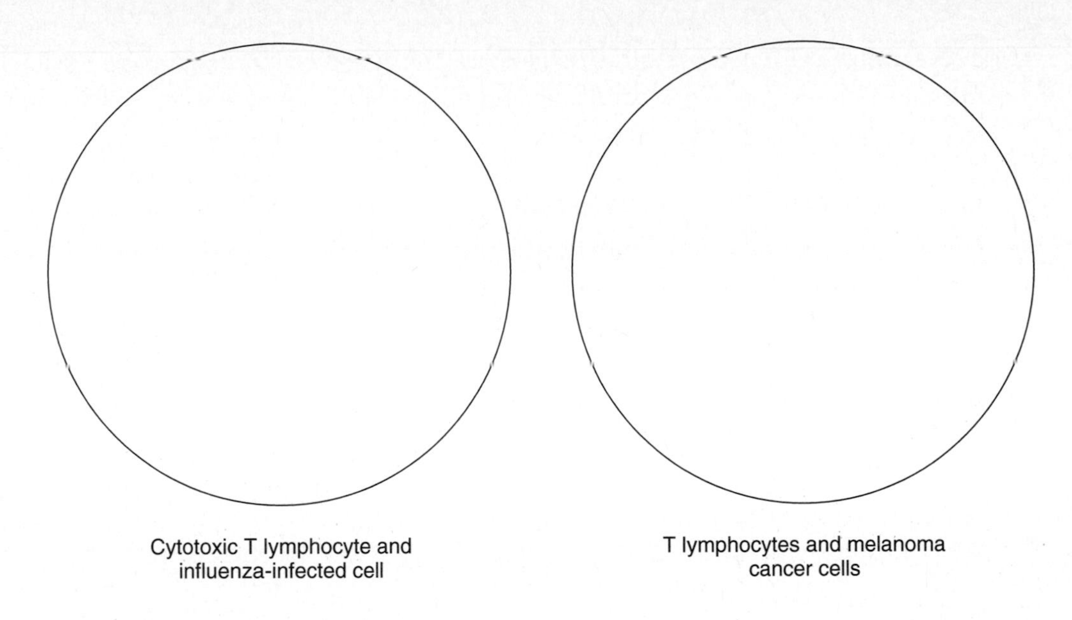

Cytotoxic T lymphocyte and influenza-infected cell

T lymphocytes and melanoma cancer cells

Label the lymphocyte and the infected cell. Label a lymphocyte and a cancer cell.

~16 min.

The last segment of the video describes vaccination to prevent disease. Epidemiologist David Fleming explains that vaccination involves injection of an antigen to stimulate a primary immune response. The memory cells that form produce antibodies if the virulent antigen is encountered at a later time. Vaccines may be (1) live, attenuated organisms, (2) dead viruses or bacteria, or (3) molecules from the outer surface of a microbe (subunits).

Video Questions

1. Discuss several reasons why we don't vaccinate against all diseases. See the box on p. 488 in the textbook.

2. Discuss what's wrong with this statement: Bacteria enter cells in order to avoid phagocytosis. The bacteria aren't planning to avoid phagocytosis; the end result isn't their goal. Falkow points out that bacteria don't have brains. More correctly, bacteria that enter cells avoid phagocytosis.

3. Why don't newborns and adults get "childhood" diseases such as measles, mumps, and chickenpox? Newborns will have passive acquired immunity and adults should have active acquired immunity that will produce a secondary immune response.

4. The video shows a photograph with the legend "Chemotaxis of neutrophils." What does this mean? White blood cells of the neutrophil type are moving toward a food source (bacteria) because of a chemical attractant.

5. In the video, Fleming says that smallpox has been eliminated. How was this accomplished? By creating herd immunity. This started when Montagu brought the Turkish practice of "ingrafting" to the attention of the Western world (see the opening quotation).

6. Fleming lists some reasons why preventable diseases are still occurring in developing countries. What are the reasons? Vaccines may not be available because of their cost or the need to refrigerate them. Public health standards are not the same as in developed countries (see Unit 9).

7. Why are some preventable diseases still occurring in developed countries (e.g., measles in the U.S., diphtheria in eastern Europe)? Many people in developed countries are complacent about vaccinations because epidemics and hundreds of deaths due to these diseases don't occur in developed countries. Left unvaccinated, a generation of susceptible children could be host to a disease.

EXERCISES

Concept Map 11.1

This map shows the relationship between the components of blood introduced in this unit. Be sure that you can identify the function of each cell type.

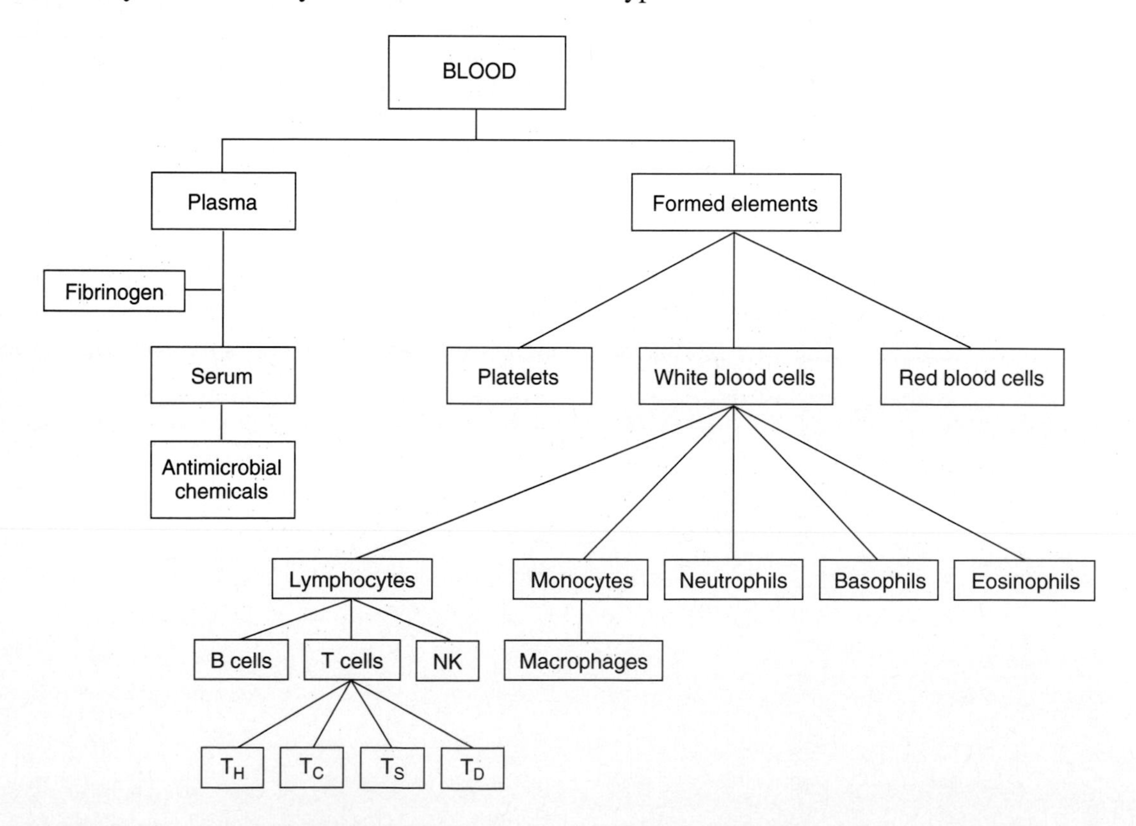

1. Name one nonspecific and one specific antimicrobial chemical in blood.
2. Which cells are shown in the video?
3. Which cells are phagocytes?
4. Which cells release histamine?
5. Which cells make antibodies?
6. Where would you put plasma cells and memory cells in this map?

Figure 11.1

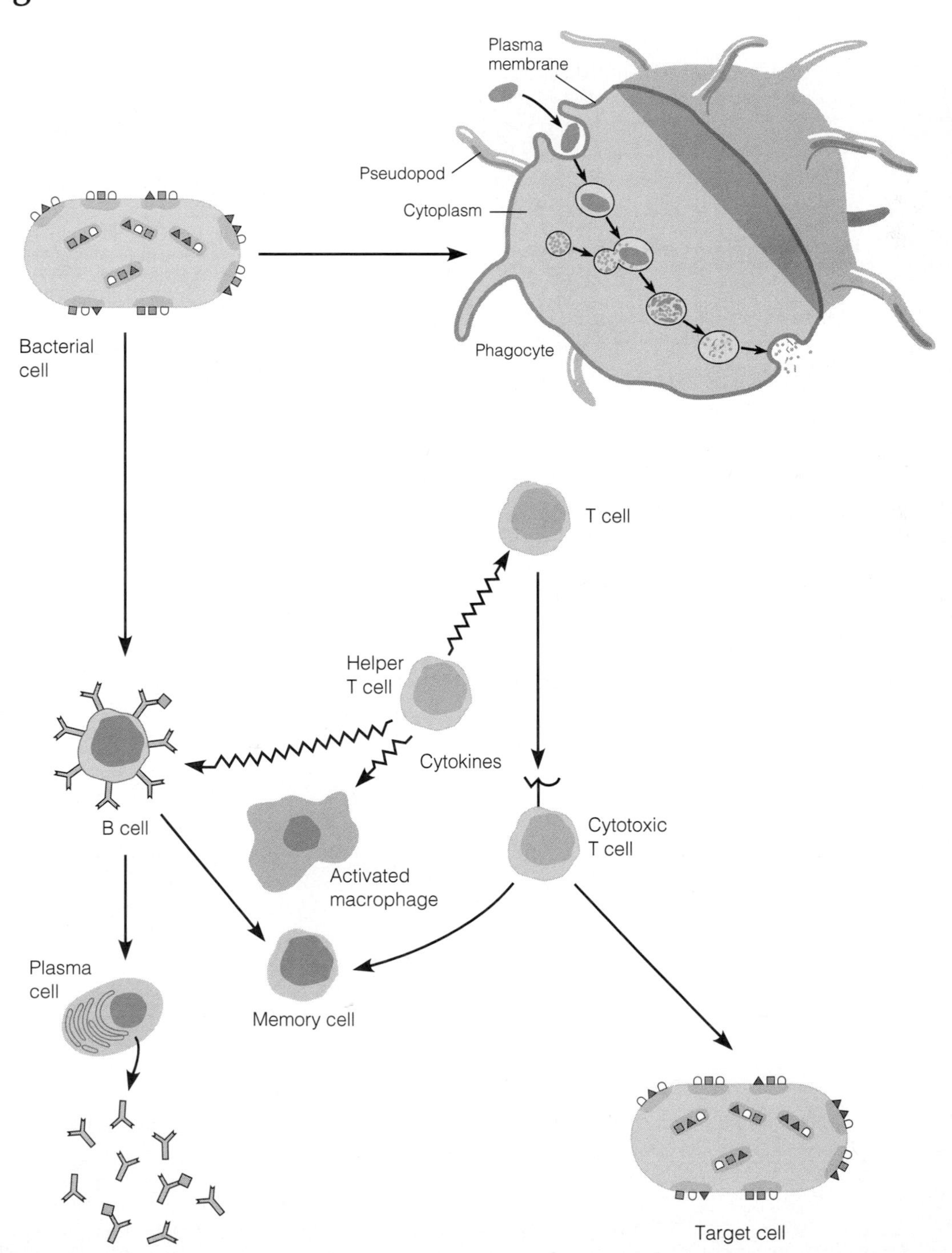

1. When does a B cell require stimulation by a T_H cell?

2. What is the function of T_C cells?

3. Where are the antigens in this diagram? The antibodies?

4. Add IL-1 and IL-2 to this figure.

Figure 11.2

The following terms can be found in Chapter 19 of your textbook. This activity will introduce you to immune deficiencies and help you understand the categories of immunity. Make a concept map showing the relationships among the following terms:

1. Acquired
2. AIDS
3. Artificial
4. Congenital
5. Cyclosporine
6. Immunodeficiency
7. Natural
8. Severe combined immunodeficiency (SCID)

Immunity Questions

Match the following choices to the statements below:

a. Artificially acquired active immunity
b. Artificially acquired passive immunity
c. Innate resistance
d. Naturally acquired active immunity
e. Naturally acquired passive immunity
f. Nonspecific resistance

1. The type of protection a person needs after being bitten by a rabid raccoon. ____
2. The type of protection a Peace Corps volunteer has after being vaccinated against rabies. ____
3. The type of protection a person has after recovering from mumps. ____
4. The type of protection an infant has at birth. ____
5. A human's protection against feline leukemia. ____
6. The protection afforded by phagocytes. ____

Humoral-Immunity Questions

Match the following choices to the statements below:

a. IgA
b. IgD
c. IgE
d. IgG
e. IgM

1. Can cross the placenta, fixes complement. ____
2. First antibody to appear, fixes complement. ____
3. Found in secretions. ____
4. Involved in allergic reactions. ____
5. On the surface of B cells. ____

Cellular-Immunity Questions

Match the following choices to the statements below:

a. B cell
b. Macrophage
c. NK
d. T_C cell
e. T_D cell
f. T_H cell
g. T_S cell

1. Acts as an antigen-presenting cell. ____
2. Secretes IL-2. ____
3. Also called CD4 cell. ____
4. Regulates the immune response when the antigen is no longer present. ____
5. Presents antigens to B cells. ____

Nonspecific-Defenses Questions

Match the following choices to the statements below:

a. Complement
b. Cytokine
c. Histamine
d. Interferon
e. Kinins
f. Prostaglandins

1. Serum proteins that cause vasodilation. ____
2. Interferes with viral replication. ____
3. Released from most cells; cause capillary dilation. ____
4. Serum proteins that can lyse cells. ____
5. Serum proteins that enhance emigration of phagocytes. ____
6. Used for communication between immune system cells. ____

Short-Answer Questions

1. Read more about the normal microbiota of the human body (pp. 397, 632, 659, and 694 in your textbook) and complete the following table:

Organ/System	Normal Microbiota	Value to human host?
Skin		
Respiratory		
Digestive		
Urogenital		

2. In a laboratory, *Staphylococcus* bacteria are isolated by inoculating a sample onto a nutrient medium containing 7.5% NaCl. Bacteria that grow on this medium are probably staphylococci or micrococci. What keeps other bacteria from growing on the medium? If you transfer the bacteria to a nutrient medium without salt, the colonies will grow much larger. Why?

3. Do people with AIDS make antibodies? Do you think the opportunistic infections that people with AIDS get are T-dependent or T-independent? Briefly explain.

4. Skin tests such as the tuberculin skin test are used to determine the presence of cell-mediated immunity against an antigen. Dispute or support the statement "It's better to be tuberculin positive." What does a seroconversion indicate?

5. Why do humans get some diseases such as chickenpox only once in their lifetime but can acquire others such as influenza dozens of times?

A Bioethics Question

Question: In the United States, 121 of the 123 cases of polio reported since 1980 were due to the vaccine. How can the polio vaccine cause polio? Discuss whether use of the polio vaccine is warranted. Use your textbook to find out how polio is transmitted.

Discussion: Polio is still occuring in a few countries, although the WHO plan is to eradicate the disease by 2000–2001. Students should note that (worldwide) herd immunity is important for preventing a reemergence of the disease. Students should also note that adequate sewage treatment will prevent epidemic transmission of polio.

Conclusion:

Study Questions

***Microbiology: An Introduction,* Sixth Edition**

Pages	Review Questions	Multiple Choice Questions	Critical Thinking Questions	Clinical Applications Questions
459–460	All	All	All	All
483–484	All	All	All	All
502	1			

CD Activities

1. Do the Chapter 16 and Chapter 17 quizzes.
2. Use the Interactive Unit on Immunology to get hands-on practice with T cells and B cells.

Web Activities

1. Do the Chapter 16 and Chapter 17 quizzes.

2. Read "The Mechanism for Antibody Diversity."

   a. Assume that the gene pool for heavy chains consists of five V, four D, and four J segments. How many different heavy chains are possible from these segments? 400

   b. Diagram the events leading to the production of a heavy chain consisting of $V_3D_2J_3C$.

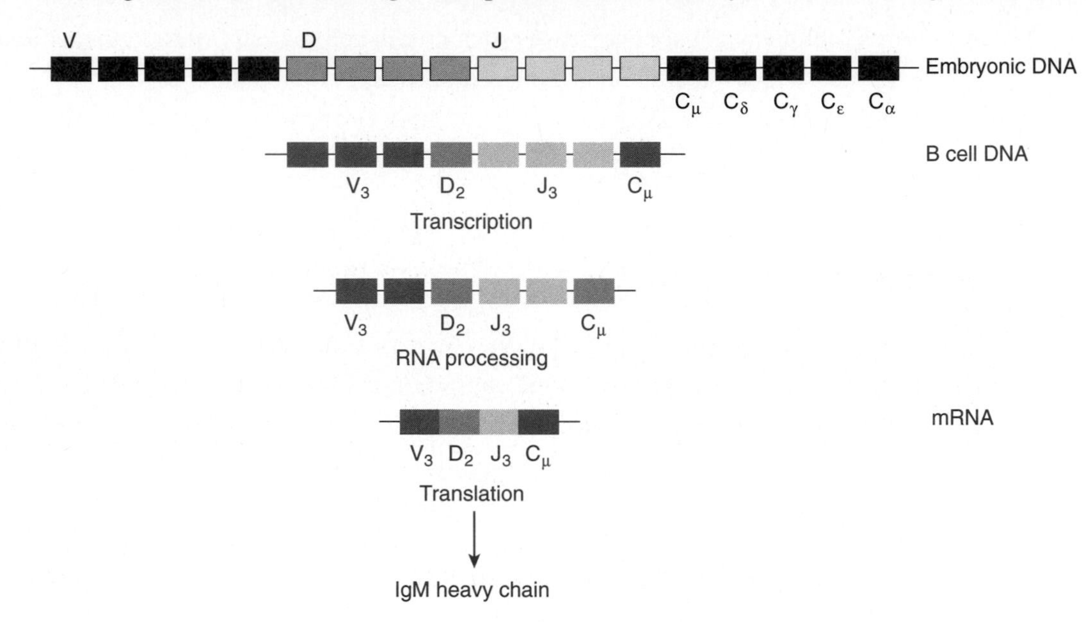

   c. Diagram the events leading to an IgM antibody with this variable region: $V_3D_2J_3$. As above plus the production of a light chain.

   d. Are all the cells in the human body genetically identical? No. Each B cell and helper T cell is genetically unique. (Red blood cells do not have DNA.)

ANSWERS

Concept Map 11.1

1. Transferrins, antibodies
2. Neutrophils, macrophages, helper T cells, cytotoxic T cells, NK cells
3. Macrophages, neutrophils, eosinophils
4. Basophils
5. Plasma cells
6. Branching from B cells

Figure 11.1

1. A B cell requires T_H cells for T-dependent antigens.
2. T_C cells lyse target cells.
3. The antigens are on the bacterial cell and then on the APC (phagocyte); see Figure 17.3 (p. 464) and Figure 17.18 (p. 481).
4. See Figure 17.12 (p. 476); the APC secretes IL-1, which activates the T_H cell to produce IL-2. IL-2 causes the T_H cell to reproduce.

Figure 11.2

Your map should start with the most general term, immunodeficiency, and end with the most specific terms. One possible map is shown here.

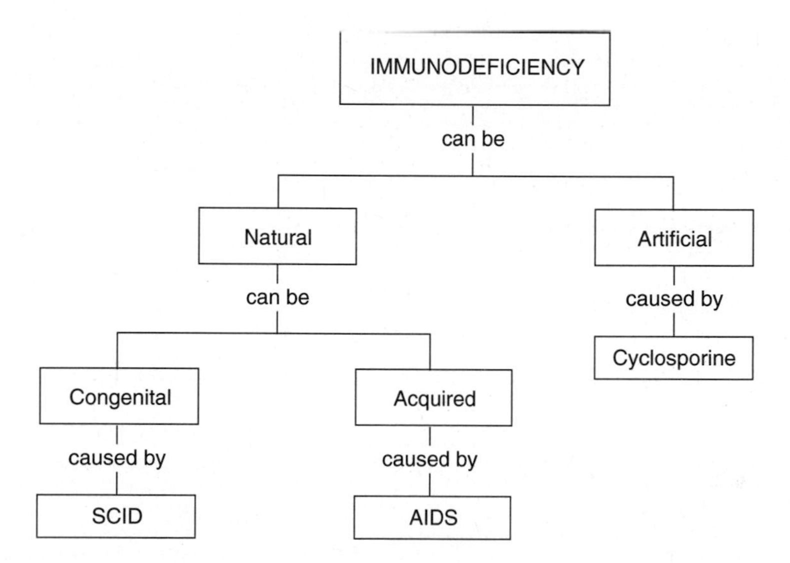

Immunity Questions

1. b
2. a
3. d
4. e
5. c
6. f

Humoral-Immunity Questions

1. d
2. e
3. a
4. c
5. b

Cellular-Immunity Questions

1. b
2. f
3. f
4. g
5. f

Nonspecific-Defenses Questions

1. e
2. d
3. c
4. a
5. f
6. b

Short-Answer Questions

1.

Organ/System	Normal Microbiota	Value to human host?
Skin	See Table 14.1 (p. 397)	Compete for food source of pathogens; bacteriocins inhibit pathogens.
Respiratory	See Table 14.1 (p. 397)	Compete for food source of pathogens; bacteriocins inhibit pathogens.
Digestive	See Table 14.1 (p. 397)	Compete for food source of pathogens; bacteriocins inhibit pathogens; *E. coli* produces folic acid and vitamin K.
Urogenital	See Table 14.1 (p. 397)	Produce bacteriocins that inhibit pathogens; produce acids that inhibit *Candida*.

2. The bacteria that have evolved to live on human skin will tolerate the hypertonic environment created by 7.5% NaCl. These bacteria from skin are not halophiles and will grow better without the salt.
3. The test to determine whether a person is HIV-positive is a test for the presence of antibodies against HIV; therefore, an HIV-positive person does make antibodies. The opportunistic infections are most likely T-dependent antigens. The HIV-positive/AIDS patient has fewer T_H cells than normal and can't respond to T-dependent antigens.
4. A positive tuberculin skin test (TB+) indicates the presence of cell-mediated immunity against *M. tuberculosis*. This immunity can result from a current infection, a vaccination, or a prior infection, causing the person to be immune to tuberculosis. If a seroconversion to TB+ occurred, the person probably has an infection.
5. The memory response can protect against the disease. Some antigens are not "good" antigens and don't cause a long-lasting response. Some pathogens like *Influenzavirus* mutate frequently, and their antigens change, so the original antibodies don't recognize them.

Unit **12**

MICROBES AND HUMAN DISEASE

Humboldt has observed, that "under the torrid zone, the smallest marshes are the most dangerous, being surrounded, as at Vera Cruz and Carthagena, with an arid and sandy soil, which raises the temperature of the ambient air" . . . In all unhealthy countries, the greatest risk is run by sleeping on shore. Is this owing to the state of the body during sleep, or to a greater abundance of miasma at such times?

—CHARLES DARWIN, 1845

LEARNING OBJECTIVES

	1. List Koch's postulates.	p. 398
	2. Differentiate between a communicable and a noncommunicable disease.	p. 400
	3. Categorize diseases according to frequency of occurrence.	p. 400
	4. Categorize diseases according to severity.	p. 401
	5. Define herd immunity.	p. 401
★	6. List five probable reasons for emerging infectious disease, and name one example for each reason.	p. 401
★	7. Define reservoir of infection.	p. 404
★	8. Contrast human, animal, and nonliving reservoirs, and give one example of each.	p. 404
	9. Explain four methods of disease transmission.	p. 406
	10. Define nosocomial infections, and explain their importance.	p. 409
	11. Define compromised host.	p. 410
	12. Identify four predisposing factors for disease.	p. 411
	13. Put the following in proper sequence, according to the pattern of disease: period of decline, period of convalescence, period of illness, prodromal period, incubation period.	p. 412
★	14. Define epidemiology, and describe three types of epidemiologic investigations.	p. 413
★	15. Identify the function of the CDC.	p. 415
	16. Define the following terms: morbidity, mortality, and notifiable disease.	p. 415

READING

Chapter 14 (pp. 398–416) and pp. 422, 617.

SUGGESTED LABS FROM JOHNSON AND CASE, *LABORATORY EXPERIMENTS IN MICROBIOLOGY*, FIFTH EDITION

Exercise 39: Epidemiology
Exercise 40: Koch's Postulates
Exercises 45–50: Microorganisms and Disease

KEY TERMS AND CONCEPTS

acute disease p. 401
analytical epidemiology p. 413
bacteremia p. 404
biological transmission p. 408
carriers p. 404
Centers for Disease Control and Prevention (CDC) p. 415
chronic disease p. 401
communicable disease p. 400
compromised host p. 410
contact transmission p. 406
contagious diseases p. 400
descriptive epidemiology p. 413
direct contact transmission p. 406
droplet transmission p. 407
emerging infectious diseases (EIDs) p. 401
endemic disease p. 400
epidemic disease p. 400
epidemiology p. 413
etiology p. 394
experimental epidemiology p. 414
focal infection p. 404
incidence p. 400
incubation period p. 412
indirect contact transmission p. 407
Koch's postulates p. 398
latent disease p. 401
local infection p. 404
mechanical transmission p. 408
MMWR p. 415
morbidity p. 415
morbidity rate p. 415
mortality p. 415
mortality rate p. 415
noncommunicable disease p. 400
nosocomial infection p. 409
notifiable diseases p. 415
period of convalescence p. 412
portals of exit p. 408
predisposing factor p. 411
prevalence p. 400
primary infection p. 404
prodromal period p. 412
reservoir of infection p. 404
secondary infection p. 404
septicemia p. 404
signs p. 400
sporadic disease p. 400
subacute disease p. 401
subclinical (inapparent) infection p. 404
symptoms p. 400
syndrome p. 400
systemic (generalized) infection p. 404
toxemia p. 404
vectors p. 408
vehicle transmission p. 407
viremia p. 404
zoonoses p. 404

INTRODUCTION

Be sure to read the Study Outline for Chapter 14, pp. 416–418, and the box on p. 422.

Although most microbes are beneficial to the ecosystem and do not cause disease, many microbes are opportunistic pathogens—that is, they don't normally have access to a host, but if put in a host when the host is weakened from another disease or malnourished, disease results. For example, people whose immune systems are weakened by AIDS or cancer chemotherapy can get infections such as *Pneumocystis* pneumonia to which healthy people are resistant. Hospitalized patients and patients undergoing surgery can acquire infections during their medical stay. These are called nosocomial infections.

A few microbes that are capable of growing in and on a host almost always harm that host. Recall from Unit 10 that these microbes are called parasites or pathogens. There is a wide range of pathogens, from the Rhinoviruses to *Vibrio cholerae* to *Mycobacterium tuberculosis*. *Rhinovirus* causes the common colds. Colds are an example of an acute disease—that is, they last a short time. The host is able to walk around, which ensures that the virus can be transmitted to a new host. A mutant *Rhinovirus* that killed its host might not have time to be transmitted. Cholera also causes an acute and often fatal disease. The violent diarrhea of cholera ensures that the *V. cholerae* bacteria get out of the sick host so they might contact a new host. *M. tuberculosis* causes a long-term or chronic disease, which will give the bacteria a host for a long time with adequate time for transmission.

The stages of disease usually follow a pattern, although the length of time for each stage and the final outcome may vary with each disease. After contact between pathogen and host, the pathogen starts to grow during the incubation period. The length of incubation periods varies widely; look at the list of various diseases and their incubation periods on p. 424 of your textbook. The first disease symptoms appear during the prodromal period; this is followed by the period of illness. The period of illness is short for acute diseases and can be quite long for chronic diseases. The period of illness is followed by the period of decline and, usually, the period of convalescence.

Until the 1880s, people didn't know what caused infectious diseases. They knew diseases were associated with stagnant water, the miasma of which Darwin wrote in 1845 in this unit's opening quote. The association of disease with mosquito-ridden slow-moving water gave rise to the name of one of those diseases—*malaria* is from Italian for "bad air." Pasteur believed that there was a relationship between microorganisms and disease because he had proven that microorganisms caused "diseases" or spoilage of food. Both Pasteur and Agostino Bassi had observed microorganisms growing on diseased silkworms. However, it isn't enough to see the microorganism on the diseased host; the microorganism could be the result of the disease and not the cause or could be a natural inhabitant of the host. It was Robert Koch who provided a procedure to prove that a specific microorganism causes a particular disease. We still use his procedure, called Koch's postulates, to determine the cause of new diseases. In recent years, Koch's postulates have been employed to discover the cause of several diseases including Lyme disease, ulcers, and Legionnaires' disease.

With the advent of antibiotics, vaccinations, and sanitation, people living in developed countries believed that infectious diseases would be a thing of the past. However, modern communications and new techniques for identifying microbes have made us more aware of emerging infectious diseases in recent years. The study of disease transmission and distribution is called epidemiology. Epidemiologists monitor the incidence and prevalence of diseases. The central source of epidemiologic data in the United States is the Centers for Disease Control and Prevention (CDC). The CDC works with health agencies around the world to monitor the spread of diseases.

VIDEO: MICROBES AND HUMAN DISEASE

Preview of Video Program

This video program introduces the science of epidemiology using emerging *Hantavirus* infections as an example. You can read more about some of the pioneers of epidemiology in your textbook: Ignaz Semmelweis (p. 9), John Snow (p. 413), and Florence Nightingale (question

2, p. 419). In the video, medical doctors and CDC epidemiologist C. J. Peters discuss epidemiologic techniques used to determine the cause of *Hantavirus* pulmonary syndrome in the American Southwest. Tissue damage resulting from *Hantavirus* infection is caused by cytokines (Unit 11).

The second example of epidemiology shows Vanya Gant, Gui Thwaits, and Mark Taverner looking through church records to determine whether English sweating sickness was caused by a *Hantavirus*.

In the third example of epidemiology, C. J. Peters worked with Delia Enria to determine the method of transmission of Andes *Hantavirus*. Peters and Enria developed several alternative hypotheses about the disease. The hypotheses are essentially questions. For example, H_1: Is the disease transmitted by direct contact with rodents? H_2: Is the disease transmitted by inhalation of rodent urine and feces? H_3: Is the disease transmitted by direct contact with humans? They tested each hypothesis to eventually rule out all except which one? human contact In the *Hantavirus* investigations, you'll see epidemiologists working at Biosafety Level 4 (Appendix C).

At ~5 min. in the video, there are time-lapse videos of a growing chain of streptococci and a macrophage moving. Draw them in the spaces below.

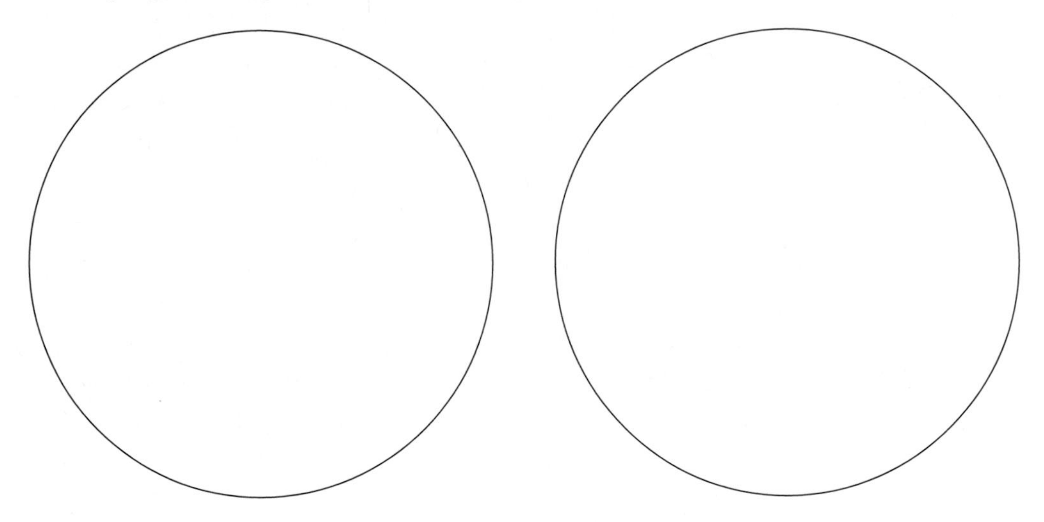

Chain of streptococci. Label one cell. Macrophage. Label the pseudopods.

Video Questions

1. In the video, you learned that each of five outbreaks of English sweating sickness stopped inexplicably. How does an epidemic stop *without* medical intervention (e.g., vaccination, chemotherapy, quarantine)? When susceptible hosts are no longer available, usually because the susceptible hosts got sick and either died or developed immunity and recovered. Some diseases like *Hantavirus* or Ebola virus might stop because the pathogen hasn't evolved an effective method of transmission or ability to grow in a new host.

2. Identify factors that contribute to the emergence of infectious disease. Migration, especially important in modern times because air travel can move a sick person quickly. Developing new areas, e.g., clearing rain forests. Natural phenomena, e.g., earthquakes, floods, increases in reservoir populations. Pollution, which may weaken the immune system.

3. Explain Jim Leinfelder's statement "Humans are not a good host for *Hantavirus* . . . yet." Use the 1996 outbreak of *Hantavirus* in Argentina to explain the difficulty viruses have crossing a species barrier. (That is, most viruses infect only one or two different host species.) The virus normally does not infect humans. We may be seeing some viruses that inadvertently came into contact with humans, but the viruses don't have the right genes to grow in a human. However, mutations are always occurring, and a virus that can grow in human cells efficiently may come in contact with the human population.

4. Do microbes "plan" to cause disease? No, microbes that penetrate our outer defenses simply need a means to feed and reproduce (Unit 11).

5. The three methods used in epidemiology are descriptive, analytical, and experimental. Match the three examples in the video to the epidemiological method used. Descriptive—English sweating sickness; analytica—*Hantavirus* pulmonary syndrome; and experimental—Andes *Hantavirus*.

6. The diseases or organisms in the table on the following page are mentioned in this video. Using your textbook, fill in the table.

Disease	Causative Agent (Genus and Species)	Bacteria, Protozoan, Virus, or Fungus	Gram Reaction, if Bacterial	Organ/ System Affected	Method of Transmission	Treatment
Meningitis or gonorrhea	*Neisseria*	B	–	CNS; urogenital	Droplet; direct contact	Antibiotics
Yeast infection	*Candida albicans*	F		Skin; mucous membranes	Endogenous; direct contact	Antifungal drugs
Salmonellosis	*Salmonella (enterica)* serotypes	B	–	Digestive	Vehicle transmission	Fluids & electrolytes
Ebola hemorrhagic fever	Ebola virus	V		Circulatory & lymphatic	Contact with blood	None
Influenza	*Influenzavirus*	V		Respiratory	Droplet	Antiviral drugs reduce symptoms
AIDS	HIV *(Lentivirus)*	V		Circulatory & lymphatic	Direct contact; parenteral	Anti-AIDS drugs reduce symptoms
Legionnaires' disease	*Legionella pneumophila*	B	–	Respiratory	Droplet	Antibiotics
Hantavirus pulmonary syndrome	*Hantavirus*	V		Respiratory	Aerosols from rodent urine	None
Plague	*Yersinia pestis*	B	–	Circulatory & lymphatic	Flea bite	Antibiotics
Systemic plague	*Yersinia pestis*	B	–	Respiratory; circulatory & lymphatic	Droplet	Antibiotics
Anthrax	*Bacillus anthracis*	B	+	Circulatory & lymphatic	Direct contact; droplet	Antibiotics

EXERCISES

Concept Map 12.1

This map shows terms related to disease and epidemiology.

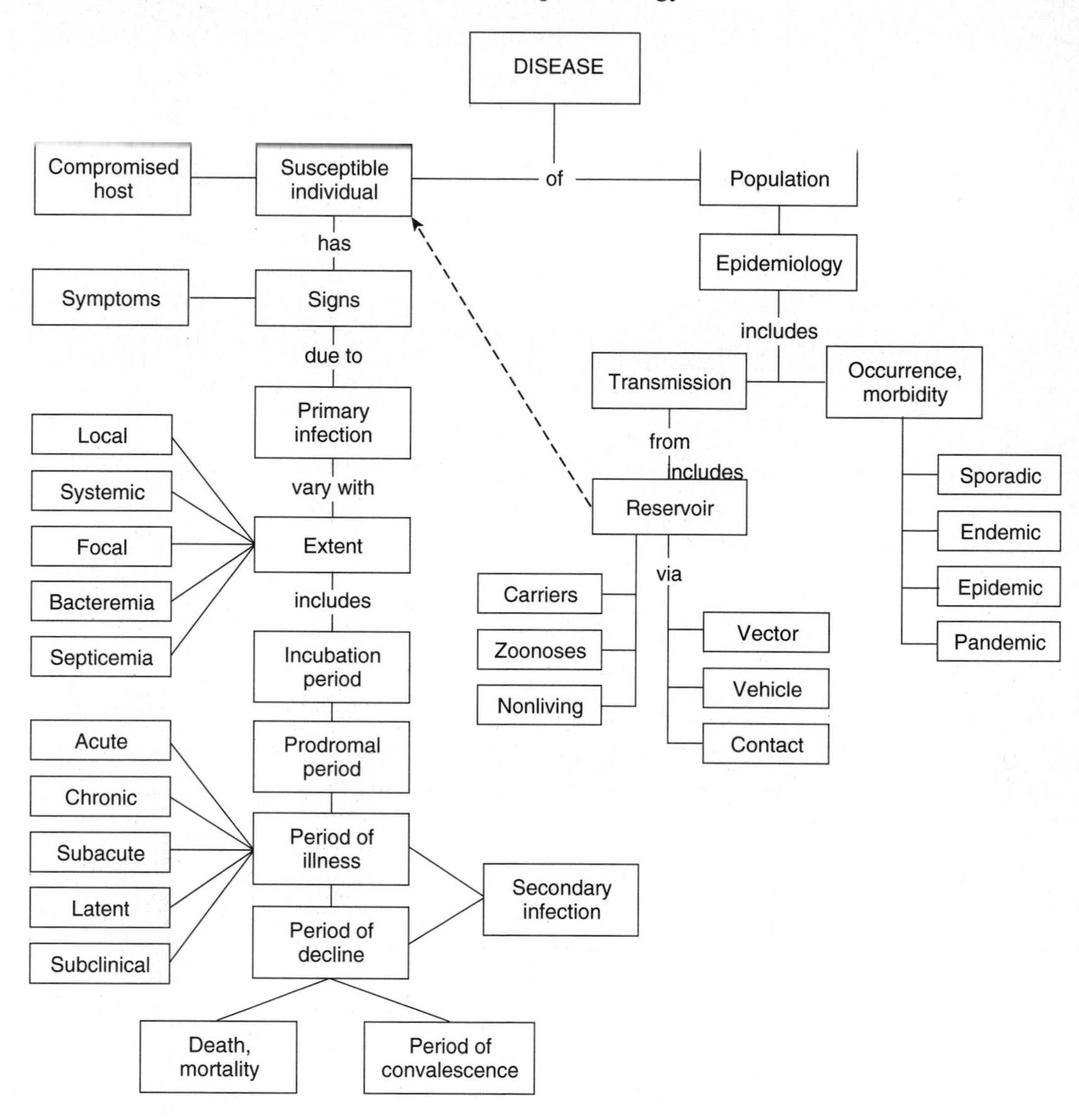

1. Where would you put the following terms?
   a. Etiology
   b. Drinking water
   c. Hypodermic needle
   d. Animal with rabies
   e. Headache
   f. Fever
   g. Housefly
   h. Blood-sucking insect
2. With which side of the map is the CDC concerned?

Figure 12.1

1. Label each step in the figure.
2. Why don't Koch's postulates work for all infectious diseases?

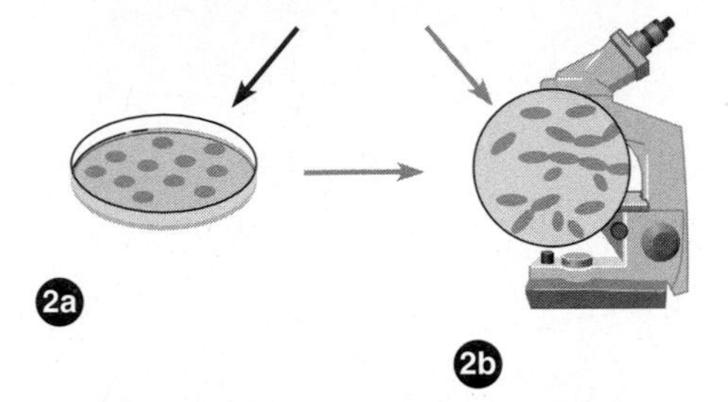

3. Why are Koch's postulates necessary? Can't you go from step 1 to step 4?

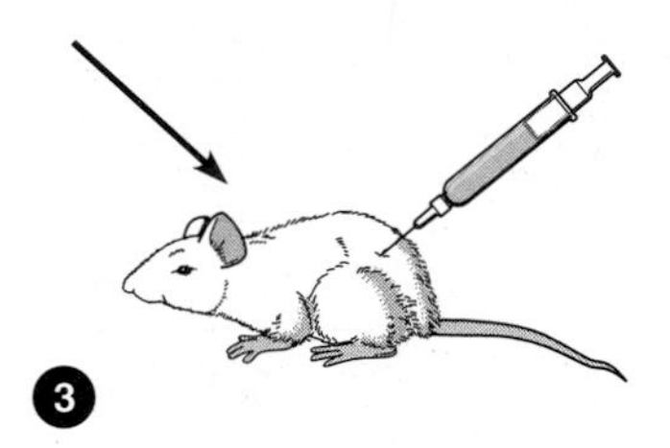

4. Will identifying the etiology of a disease prevent spread of the disease? Discuss why or why not

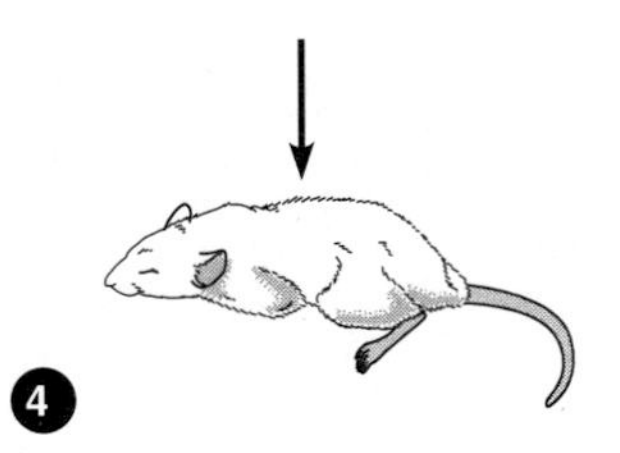

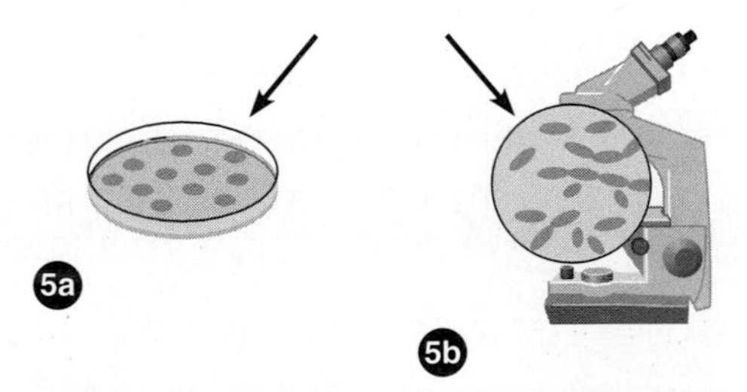

Definitions

Match the following statements to words from Key Terms and Concepts:

1. Time between contact with a pathogen and first symptoms. ____________________
2. Disease that is not spread from one host to another. ____________________
3. An infection acquired in the hospital. ____________________

4. The study of the cause of a disease. ______________________
5. Blood-sucking insects. ______________________
6. Comparison of a healthy population with a diseased population. ______________________
7. Person whose resistance to infection is impaired. ______________________
8. Infection caused by an opportunistic microbe during another infection. ______________________
9. An infection throughout the body. ______________________
10. Disease that develops rapidly and lasts a short time. ______________________

An Epidemiologic Problem

An infestation of *Aedes albopictus* (Asian tiger mosquito), a mosquito known to transmit epidemic dengue fever in its native Asia, was discovered in 1985 in Harris County, Texas. The mosquito was probably introduced in standing water in used tire casings imported from Asia. The mosquito is a competent vector for the yellow fever and dengue fever viruses. The maps show (a) the distribution of *A. albopictus* in the United States, and the endemic areas for (b) dengue fever and (c) yellow fever in the Americas.

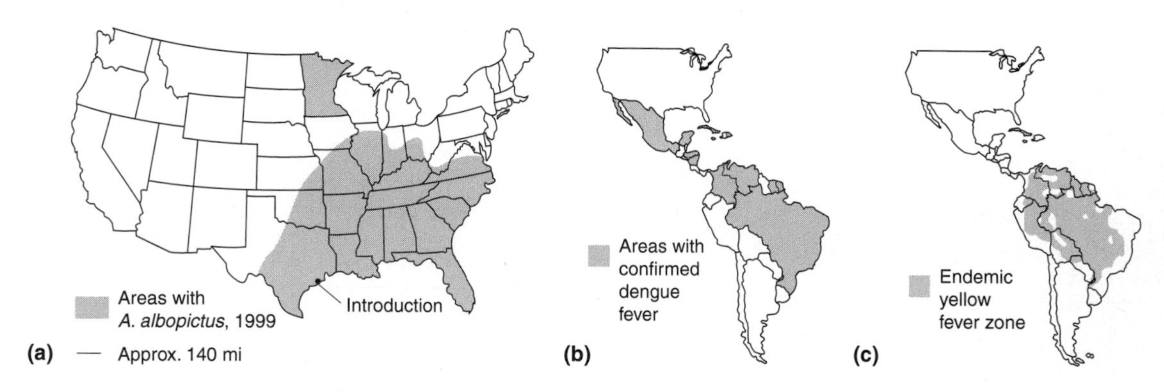

1. Estimate the rate of movement of *A. albopictus* in the United States. ~70 miles/year.

2. Will the northward movement of this mosquito ever stop? Why or why not? Yes, when the mosquito reaches the lowest temperatures at which it can live.

3. How long will it take *A. albopictus* to meet the dengue fever endemic areas? They have met at the U.S.-Mexico border.

4. How long will it take *A. albopictus* to meet the yellow fever endemic areas? (1 cm ≈ 2000 mi.) It could take 20 years by land or just a few hours by plane across the Caribbean Sea.

5. What are the normal vectors for these diseases in the United States? (*Hint:* Look them up in your textbook, Chapters 12 and 23.) *Aedes aegyptii.*

6. How can you account for the discontinuous infestation in Minnesota? The mosquitoes probably hitched a ride on a truck or train.

7. What could be the potential consequences of introduction of *A. albopictus* in the United States? Transmission of dengue fever in the United States. The mosquito may also be a suitable vector for other viral diseases, including Eastern equine encephalitis.

8. Suggest methods of controlling the mosquito population. Mosquito repellent, mosquito netting, eliminating artificial breeding areas such as buckets that fill with rainwater.

Fill-In Questions

Use the following choices to complete the statements below. Choices may be used once, more than once, or not at all.

carrier	latent disease	subacute
endemic	local infection	subclinical infection
epidemic	mechanical transmission	systemic infection
fomite	period of illness	vehicle transmission
gram-negative enterobacteria	prodromal period	
gram-positive cocci	secondary infections	

1. Cockroaches have not been associated with disease transmission. If one did transmit a disease, it would be by ______________________.
2. A pimple is an example of a(n) ______________________.
3. The etiology of most nosocomial infections is ______________________.
4. A disease that is constantly present in a population is ______________________.
5. A rise in prevalence of a disease is a(n) ______________________.
6. Opportunistic pathogens cause ______________________ in AIDS patients.
7. Contaminated food transmits disease by ______________________.
8. A disease with onset time between acute and chronic is ______________________.
9. The time of the first mild symptoms of a disease is the ______________________.
10. Mary Mallon infected twenty-two people for whom she cooked with typhoid fever. The epidemiologist who tracked her down in 1907 was Sara Baker. Mary can be described as a(n) ______________________.

An Epidemic

During a 6-month period, 239 cases of pneumonia occurred in a town of 300 people. A clinical case was defined as fever ≥ 39°C lasting > 2 days with three or more symptoms (i.e., chills, sweats, severe headache, cough, aching muscles/joints, fatigue, or feeling ill). A laboratory-confirmed case was defined as a positive result for antibodies against *Coxiella burnetii*. Before the outbreak, 2000 sheep were kept northwest of the town. Of the 20 sheep tested from the flock, 15 were positive for *C. burnetii* antibodies. Wind blew from the northwest, and rainfall was 0.5 cm compared with 7 to 10 cm during each of the previous three years.

1. This is an example of
   a. human reservoirs.
   * b. a zoonosis.
   c. a nonliving reservoir.
   d. a vector.
   e. a focal infection.

2. The etiologic agent of this disease was
   a. sheep.
   b. soil.
   * c. *C. burnetti.*
   d. pneumonia.
   e. wind.

3. The method of transmission of this disease was
   a. direct contact transmission.
   * b. droplet transmission.
   c. indirect contact transmission.
   d. vector-borne.
   e. vehicle transmission.

4. What is the disease? Q fever

Short-Answer Questions

1. Identify the reservoir for the following diseases, and indicate whether the diseases are acute, chronic, or latent.

   a. *Hantavirus* pulmonary syndrome

   b. AIDS

   c. Q fever

2. Why are most of the "untreatable" new diseases caused by viruses? (*Hint:* Review Unit 2.)

3. Read the case history on p. 403 of your textbook. What was the incubation period for the disease? What type of transmission occurred?

A Bioethics Question

This question provides an opportunity for your students to apply their knowledge to a real public health problem and look at the complex relationships among disease, technology, and education.

Question: In Unit 9, you learned that 50,000 people die each day from infectious diseases worldwide. What is the best way to decrease this death rate?

Data: In Unit 9, Germain Hanquet gave the following reasons for the prevalence of infectious diseases in developing countries: contaminated water, lack of education, and poor hygiene. Some additional information is provided to help you. The United States is used as an example of a developed country, and Somalia as an example of a less developed country.

	United States	Somalia
Literacy rate, %	85	24
Population with access to safe drinking water, %	100	35
Death rate per 1000 population	9.2	18.1
Per capita gross national product (US$)	28,495	110
Daily food consumption, % of recommended minimum	136	67

Data source: World Health Organization

Conclusion:

Study Questions

Microbiology: An Introduction, Sixth Edition

Pages	Review Questions	Multiple Choice Questions	Critical Thinking Questions	Clinical Applications Questions
418–420	1, 3–15	All	All	All
741		5		

CD Activity

Do the Chapter 14 quiz.

Web Activities

1. Do the Chapter 14 quiz.
2. Read "Are Cat Owners and Homeless at Risk?"
   a. How was the identity of *Bartonella* established if the bacteria couldn't be cultured? PCR was used to amplify DNA in skin lesions. Primers for genes for bacterial rRNA were used. The DNA sequences of the amplified rRNA genes were compared to known sequences. (See Unit 7.)

   b. Why don't more cat owners get *Bartonella* infections? The nonspecific defenses of normal, healthy people are probably sufficient to prevent infection.

ANSWERS

Concept Map 12.1

1. a. Perhaps along the vertical lines from the individual and population because the map is following the etiologic agent.
   b. Vehicle transmission
   c. Fomite, indirect contact transmission
   d. Zoonoses, contact transmission
   e. Symptom
   f. Sign
   g. Vector, mechanical transmission
   h. Vector, biological transmission
2. Population

Figure 12.1

1. Check your answers with Figure 14.3 (p. 399) in your textbook.
2. Some microbes, especially viruses, can't be grown on artificial media.
3. Koch's postulates are designed to have the microbe in contact with the test animal, not the original sick animal. You need to show that microbial growth is the *cause* and not the result of the disease.
4. Identifying the etiology alone will not change the behavior that transmits a disease. For example, AIDS is still being spread among people who know that it is caused by HIV.

Definitions

1. incubation period
2. noncommunicable disease

3. nosocomial infection
4. etiology
5. vectors
6. analytical epidemiology
7. compromised host
8. secondary infection
9. systemic or generalized infection
10. acute disease

An Epidemiologic Problem

Your instructor can give you more information on this problem.

Fill-In Questions

1. mechanical transmission
2. local infection
3. gram-negative enterobacteria
4. endemic
5. epidemic
6. secondary infections
7. vehicle transmission
8. subacute
9. prodromal period
10. carrier

An Epidemic

Your instructor can give you more information on this problem.

Short-Answer Questions

1. a. *Hantavirus* pulmonary syndrome: Acute; deer-mouse reservoir
   b. AIDS: Chronic; human reservoir
   c. Q fever: Subclinical; cattle, sheep reservoir
2. Antibiotics are not effective against viruses, and at present there are very few antiviral drugs. Recall from Unit 2 that viruses are using the host cell's metabolic machinery, so a drug would be likely to kill the host cell.
3. It looks like the incubation period was ~10 days if a person left camp on Aug. 4 and showed symptoms on Aug. 14. Vehicle transmission.

Appendix **A**

IDENTIFYING MICROORGANISMS*

LEARNING OBJECTIVE

Use a dichotomous key.

REFERENCES

Chapter 11 and Appendix A in *Microbiology: An Introduction*, Sixth Edition.

BACKGROUND

As you learned in Units 3 and 6, microbiologists are often required to identify microorganisms. Identification of bacteria usually begins with a Gram stain, and then appropriate biochemical tests are selected. You can see from the key in Figure 10.8 (p. 281) in your textbook that the same tests are not required for all organisms and the results from one test will dictate what the next test should be. Microbiologists use dichotomous keys like that in Figure 10.8 and information from *Bergey's Manual* to identify bacteria.

Usually, a microbiologist has an unknown bacterium such as Giovannoni did in Unit 1 that needs to be identified. In this exercise, you will have the identity of the bacteria and need to match the bacteria to their characteristics. A dichotomous key showing some of the characteristics of bacteria mentioned in *Unseen Life on Earth* is on the next page. Your assignment is to write each bacterial genus in the appropriate place in the key. You can get information on the bacteria named in the video series from your textbook.

The following bacteria were named in *Unseen Life on Earth:*

k. *Agrobacterium tumefaciens* (Unit 5)
a. *Bacillus anthracis* (Unit 12)
c. *Corynebacterium diphtheriae* (Unit 5)
e. *Enterococcus faecalis* (Unit 1)
m. *Escherichia coli* (Units 2, 7)
j. *Hemophilus* (Unit 5)
l. *Legionella pneumophila* (Unit 12)
b. *Mycobacterium tuberculosis* (Unit 9)
h. *Neisseria* (Units 5, 12)
i. *Oceanospirillum* (Unit 1)

*From *Unseen Life on Earth.*

p. *Proteus* (Unit 7)
o. *Salmonella* (Units 2, 9, 11, 12)
d. *Staphylococcus aureus* (Units 1, 2, 4, 11)
f. *Streptococcus pneumoniae* (Units 1, 5)
g. *Thermotoga maritima* (Unit 6)
n. *Vibrio cholerae* (Units 1, 9)
q. *Yersinia pestis* (Unit 12)

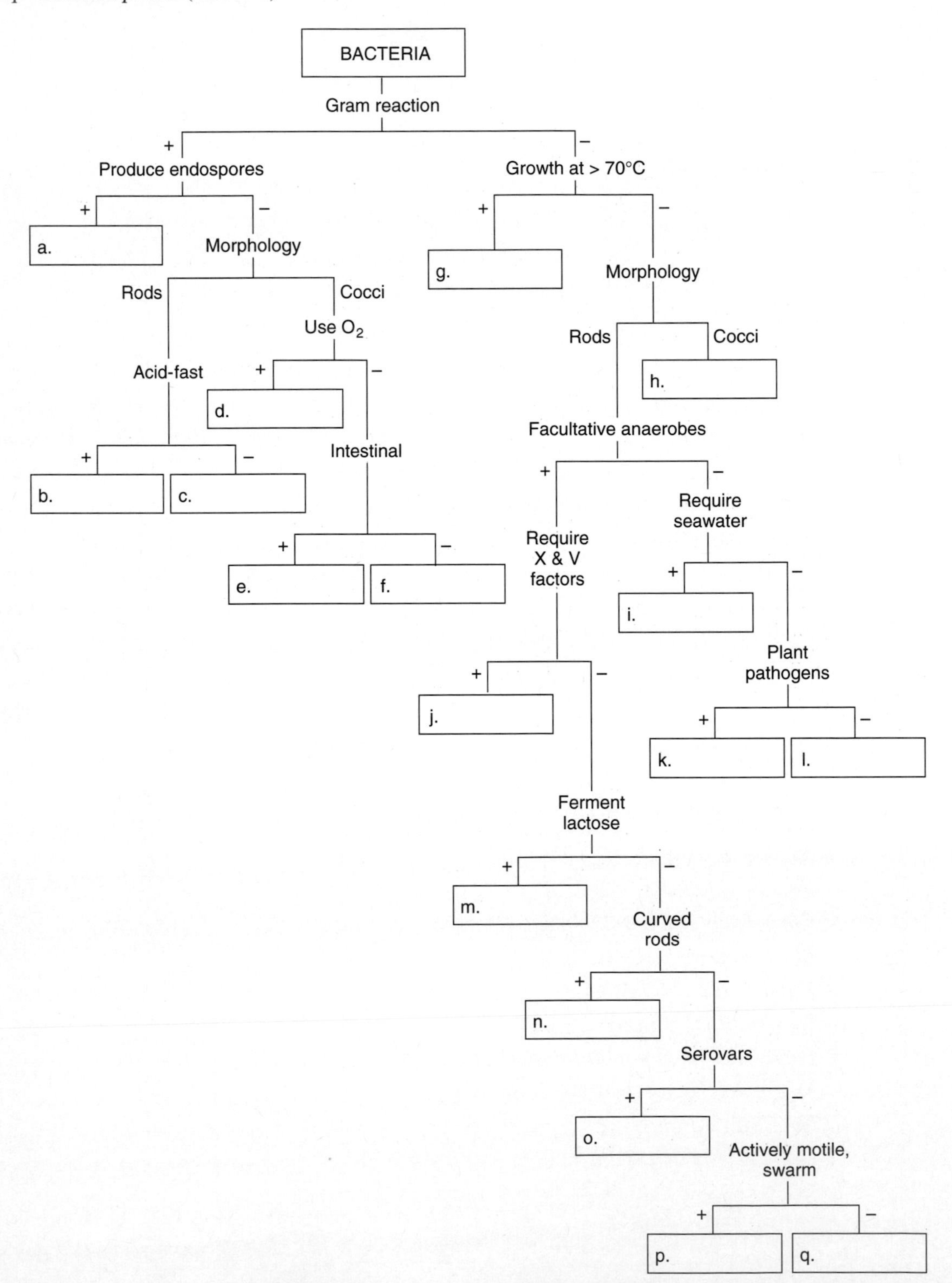

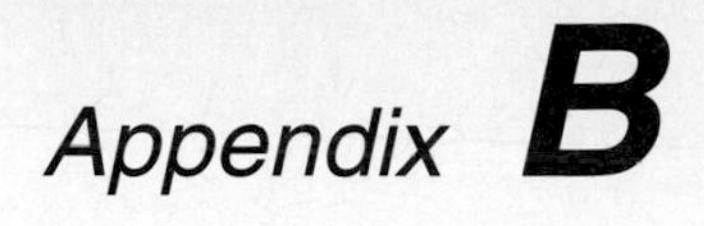

SOME FERMENTED FOODS

LEARNING OBJECTIVES

1. Define fermentation.
2. Explain how the activities of microorganisms are used to preserve food.
3. Make an enjoyable product.

REFERENCES

Pages 746–749 in your textbook; Unit 3 in *Unseen Life on Earth.*

Students could bring microbial foods for an end-of-semester party. Some foods can be prepared in a teaching lab to demonstrate fermentation and to reinforce use of metric measures. The students can also try the recipes at home. Exact measures are not necessary to make these foods. You can provide the following conversions to standard kitchen measures if your students are doing this at home:

100 ml ≈ 1/2 cup	950 ml ≈ 1 qt.	1 tbs. ≈ 14 g sugar
2.5 ml = 1/2 tsp.	1 tsp. ≈ 5 g sugar or 3 g dry milk	

Background

Microbial fermentations are used to produce a wide variety of foods. No doubt, the earliest production of these fermented foods was accidental. However, people learned that fermented foods lasted longer than the unfermented raw material. Acids and alcohols produced by fermentation inhibit the growth of acid- or alcohol-intolerant microorganisms. Additionally, bacteriocins produced by lactobacilli are known to inhibit growth of selected bacteria.

Some traditional microbial foods are listed below, and selected recipes are on the following pages. Try some of the recipes at home.

Snacks

Olives (fermented by *Leuconostoc*)
Mushrooms (fleshy fungi)
Cheeses (fermented by lactic acid bacteria and *Penicillium*)
Breads (fermented by *Saccharomyces cerevisiae* and *Lactobacillus*)

Soup

Miso (fermented by *Aspergillus* and *Saccharomyces*)

Entree

Sausages (fermented by *Pediococcus* and *Penicillium*)
Sauerkraut (fermented by *Lactobacillus plantarum*)

Dessert

Chocolate (fermented by *Kluyveromyces* and lactic acid bacteria)

Beverages

Coffee (berries fermented by *Erwinia*)
Wine (fermented by *Saccharomyces*)
Beer (fermented by *Saccharomyces*)

Yogurt

Materials

Pasteurized milk
Food thermometer
Styrofoam cup with lid
Nonfat dry milk
Yogurt starter culture
Jam or other flavoring (optional)

Procedure

1. Add 100 ml of milk per person to a wet pot (wash out pot first with water to decrease sticking of the milk).
2. Heat milk on stove to about 80°C for 40 min. Stir occasionally. Do not let it boil.
3. Cool to about 65°C and add 3 g nonfat dry milk per person. Stir to dissolve.
4. Rapidly cool to about 45°C. Pour milk equally into paper or Styrofoam cups.
5. Inoculate each cup with 2.5 ml of yogurt starter culture or 1–2 teaspoons of commercial yogurt. Cover and label.
6. Incubate at 45°C for 4–18 hours or until firm (custardlike).
7. Cool yogurt to about 5°C, then taste with a clean spoon. Add jam or some other flavoring.

Root Beer

Materials

Root beer extract*
Sucrose
Baker's yeast
Bottles, washed and sterilized
Corks or caps and crowner
Raisins (optional)

*Available at wine-making supply stores or
Zatarain's, Inc., New Orleans, LA 70114, or
McCormick/Schilling,1-800-474-7742.

Procedure

1. Pour 2.5 ml root beer extract over 64 g sugar in a kettle. Mix well.
2. Add lukewarm water to equal 950 ml.
3. Mix 0.5 g dried yeast with 5 ml lukewarm water. Let stand for 5 min. Add 0.5 g dried yeast when the temperature is ~21°C (room temperature).
4. Add yeast to the sugar-extract mixture, mix, and pour into bottles immediately. A raisin may be added if you like a great deal of carbonation. Fill to within 1 cm of the top.
5. Cork securely or seal with a capper.
6. Place bottles on their sides at room temperature for 5–7 days.
7. Refrigerate before drinking.

Beer

Materials

Malt extract*
Hops*
Sugar
Water
Hydrometer*
Yeast*

*Available from beer and wine supply stores.

Procedure

1. Dissolve 250 g sugar and 500 g malt extract in 1250 ml water.
2. Boil 30 g hops in another 1250 ml water for 10 minutes. Strain the hops, pouring the liquid onto the malt-sugar solution (wort).
3. Boil the hops a second time in another 1250 ml water. Strain and add the hops liquid to the wort.

4. Boil the hops a third time in 1250 ml water and add the liquid to the wort. Cover the wort and let it cool.
5. The starting specific gravity should be 1.036–1.040. When the wort has cooled to room temperature, add 5 g of yeast.
6. Cover with a fermentation lock and ferment for 3 days.
7. Skim off the scum and continue fermenting to a specific gravity of ≤ 1.004. Strain the beer into twelve clean, crown-cap beer bottles, filling to 2–2.5 cm of the top.
8. Add 2.5 g sugar to each bottle and cap the bottles.
9. The beer clears in a week as the (lager) yeast settle on the bottom. In 2 weeks, the beer is ready for drinking.

Sauerkraut

Materials

1-liter widemouthed jar with stopper
Cabbage, 1 head
Table salt solution, 3%

Procedure

1. Chop cabbage into small pieces and pack into a clean glass jar.
2. Fill the jar with a 3% salt solution so the liquid overflows when the stopper is replaced.
3. Incubate the cabbage mixture for 5–7 days at room temperature. Press down daily to keep the cabbage covered by liquid at all times.

Kimchee

Materials

1-liter widemouthed jar with stopper
Cabbage, 1 head
Garlic, 2 cloves
Chopped hot, red chili pepper, 1
Noniodized or rock salt, 66 g

Procedure

1. Alternate layers of cabbage, garlic, pepper, and a sprinkling of salt in the jar. Press each layer down firmly until the bottle is packed full.
2. Cover the jar. During the next hour, open the jar and press down on the contents.
3. Incubate the cabbage mixture for 5–7 days at room temperature. Press down daily to keep the cabbage covered by liquid at all times.

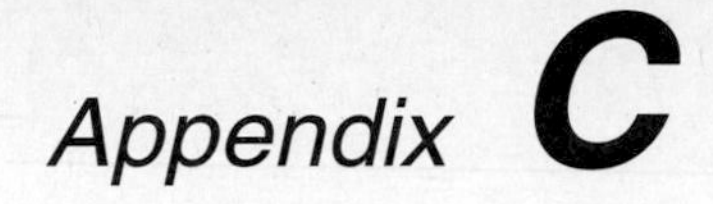

Appendix C

BIOSAFETY LEVELS*

Reference

Appendix B in your textbook.

Background

There are four biosafety levels (BSLs). Each BSL represents those conditions under which an organism can ordinarily be safely handled. You saw BSL 4 in Units 9 and 12; however, most microbiologists work at BSL 1 and 2, as shown in Units 3 and 4.

Aseptic techniques are strictly adhered to when working with any microorganisms because many microorganisms not ordinarily associated with disease processes in humans are opportunistic pathogens and may cause infection in the young, the aged, and immunodeficient or immunosuppressed individuals. Additionally, Universal Precautions are used by health care workers when in contact with patients and human body fluids. Read Universal Precautions in Appendix B in *Microbiology: An Introduction*, Sixth Edition.

Biosafety Level 1

Practices, safety equipment, and facilities are appropriate for undergraduate and secondary educational training and research laboratories. Work is done with defined and characterized strains of viable microorganisms not known to cause disease in healthy adult humans. No special barriers, other than a sink for handwashing.

Biosafety Level 2

Practices, equipment, and facilities are applicable to clinical, diagnostic, and teaching laboratories in which work is done with the broad spectrum of indigenous moderate-risk agents present in the community and associated with human disease of varying severity. With good microbiological techniques, these agents can be used safely in activities conducted on the open bench. BSL 2 is appropriate when work is done with any human-derived blood, body fluids, or tissues where the presence of an infectious agent may be unknown. Safety equipment includes splash shields, face protection, gowns, and gloves.

*Recommendations from the CDC/NIH, *Biosafety in Microbiological and Biomedical Laboratories*.

Biosafety Level 3

Practices, safety equipment, and facilities are applicable to clinical, diagnostic, teaching, research, or production facilities in which work is done with indigenous or exotic agents with a potential for respiratory transmission, and that may cause serious and potentially lethal infection. All laboratory manipulations should be performed in an enclosed chamber; there is controlled access to the laboratory and a specialized ventilation system.

Biosafety Level 4

Practices, safety equipment, and facilities are applicable for work with dangerous and exotic agents that pose a high individual risk of life-threatening disease, that may be transmitted via the aerosol route, and for which there is no available vaccine or therapy. Laboratory workers wear full-body, air-supplied positive-pressure personnel suits. The laboratory is a separate building with complex, specialized ventilation and waste-management systems.

Appendix D

CROSSWORD PUZZLE

1 H	2 I	3 V		4 L	5 A	6 G		7 G	8 R	9 A	10 M	11 S		12 P	13 C	14 R
15 F	A	D		16 I	C	E		17 M	O	N	A	S		18 I	U	D
19 R	A	R	20 E		21 E	N	22 E		23 S	U	N		24 A	L	E	S
		25 L	D		26 A	E	R		27 E	S	T		28 T	I		
29 N	30 A		31 E	32 N	E	R	G	33 Y			34 I	35 S	O		36 C	37 A
38 A	R	39 O	M	A		40 A	S	E		41 A	S	E	P	42 S	I	S
43 D	E	C	A	N	44 T			45 A	46 P	I	S		47 Y	O	L	K
		48 U		49 B	A	50 G		51 S	O	M	A	52 S		53 N	I	
54 H	55 E	L	56 D		57 P	E	58 S	T	I	S		59 M	60 N		61 A	62 H
63 Y	E	A	R	64 S		65 M	O				66 K	U	D	67 U		68 O
69 P		70 R	O	U	71 S				72 A	73 L		74 T	U	B	75 E	S
76 O	77 E		78 P	I		79 L	80 I	81 S	T	E	82 R		83 S	A	L	T
	84 B	85 R		86 S	87 T	Y	L	E		88 D	I	89 C		90 N		
91 D	O	N	92 E		93 H	E	L	P			94 P	H	95 A	G	96 O	97 S
98 P	L	A	S	99 M	I	D		100 A	101 L	102 S		103 A	M	I	N	O
104 T	A		105 T	G	A			106 L	I	P	107 A	S	E		108 E	S
		109 F	E		110 M	111 U	112 S		113 C	O	D		114 B	115 E		
116 A	117 G	A	R		118 I	R	K		119 E	R	A		120 A	M	121 E	122 S
123 W	B	C		124 S	N	A	I	125 L		126 E	P	127 I		128 I	G	E
129 E	S	E		130 D	E	L	T	A		131 S	T	D		132 T	O	M

Across

1 AIDS etiology
4 Period of enzyme synthesis
7 His method groups bacteria
12 Gene-cloning technique
15 Coenzyme in redox reactions
16 Mineral made of H_2O
17 Unit, suffix
18 Contraceptive device
19 Lanthanide series
21 Unsaturated, suffix
23 Our nearest star
24 Order suffix
25 Measure of toxicity
26 In the presence of air, prefix
27 At noon in San Francisco, it's 3 P.M. here
28 *Agrobacterium* plasmid
29 Reactive metal of table salt
31 Ability to do work
34 Same, prefix
36 Bone element
38 Odor
40 Enzyme suffix
41 Microbiologist's goal
43 Pour wine

45 Honeybee
47 Embryonic food
48 235 is a fissionable form
49 Ascus
51 Of bodies
53 Atom with 28 protons
54 Kept
57 Plague species
59 Element required by some enzyme systems
61 Exclamation
63 Duration of a slow virus infection
65 Trace element for nitrogen fixation
66 African artiodactyl
68 Gaseous element; oxidizing agent
69 A macronutrient; see 90 across
70 Showed cancer is transmissible
72 Lightweight metal
74 Pipes
76 Diphthong omitted from 95 down
78 3.14159
79 Father of aseptic surgery
83 Made from an acid and a base
84 One of the halogens
86 Digestive structure in bivalves
88 3-D microscope
90 Another macronutrient; see 69 across
91 Finished
93 A T_H cell for a B cell
94 Macrophages and neutrophils
98 Extrachromosomal DNA
100 Degenerative disease of motor neurons
103 —NH_2
104 Tantalum
105 DNA code for threonine
106 Hydrolyzes fats
108 Synthetic element
109 Element in hemoglobin and cytochromes
110 Typhus reservoir
113 *Gadus*
114 Rare metal used in X-ray tubes
116 Culture medium polysaccharide
118 Vex and bore
119 Geological time period
120 Test for mutagens
123 Monocyte or basophil
124 *Helix*
126 Goes with demic or dermis
128 Antibody on B cell
129 Compass direction
130 RNA virus that requires HBV
131 *Chlamydia*, for example
132 Male cat or turkey

Down

1 Integrated F factor
2 Natural auxin
3 Rapid detection for syphilis
4 Treatment for manic depression
5 Family suffix
6 Between families and species
7 Unit of wt.
8 Family with 5 petals and 10+ stamen
9 Exit for *Salmonella*
10 Decimal part of a log
11 Disulfide bonds form here
12 For bacterial conjugation
13 Hint
14 Nutrition experts
20 Accumulation of fluid
22 Units of 31 across
24 Spontaneous development of an allergy
29 Coenzyme in redox reactions
30 100 square meters
32 Hepatitis C
33 Unicellular ascomycete
35 Element 34
36 Motility for *Balantidium*
37 Query
39 Lens nearest the eye
41 Directs
42 Male child
44 Withdraw fluid
46 Fermented taro root
50 Prized object
52 Fungus disease of grains
54 Below, prefix
55 Virus of horse's brain
56 What body temperature does during crisis
58 For this reason
60 Neodymium + 48 across + 71 down
62 Organism harboring a pathogen
64 Species of *Brucella*
66 Potassium
67 West African tribe
71 A macronutrient
72 Heaviest halogen
73 Guided
75 Elevated railroad
77 A filovirus
79 Combined with NaOH
80 Not well
81 Calyx part
82 Strong seaward current
85 One of the nucleic acids

87 A B vitamin
89 Provided mechanism for the theory of evol.
91 Vaccine for children
92 R′COOR
95 It uses pseudopods
96 Single
97 ••• - - - •••
99 Trace element for DNase
101 Vectors for epidemic typhus
102 These can be sexual or asexual
107 Basis for natural selection
109 Consisting of eyes, nose, and mouth
111 Russian mountains
112 Short satire
115 Give off
116 Precedes some and struck
117 Neurologic side effect
121 I
122 An electron microscope
124 Pierre is the capital
125 Rare earths start here
127 Amount needed to cause infection